Thermodynamics and Statistical Physics

A Series of Books in Physics

EDITORS:
Henry M. Foley
Malvin A. Ruderman

Thermodynamics and
Statistical Physics

A SHORT INTRODUCTION

Robert J. Finkelstein

UNIVERSITY OF CALIFORNIA
LOS ANGELES

W. H. FREEMAN AND COMPANY
SAN FRANCISCO

Library of Congress Catalog Card Number: 78-94103
Standard Book Number: 7167 0325-4
Printed in the United States of America

Contents

3. One-component Systems, Phase Transitions, and Low Temperatures

4. The Third Law and the Behavior of Matter Near Absolute Zero

5. General Conditions of Thermodynamic Equilibrium and Other Applications

6. Statistical Foundations of Thermodynamics

7. Applications to Some Simple Systems

8. Thermodynamics of Ideal Quantum Gases

9.   The Grand Partition Function and Second Quantization

10.   Phase Transitions in Statistical Mechanics

11.   The Approach to Thermodynamic Equilibrium

Preface

This book is concerned with the properties of matter in bulk or of physical systems with many degrees of freedom. It was written from the point of view of the nonspecialist, and is based on a graduate course, the first of a three-course sequence that has been offered at the University of California at Los Angeles. The two courses following in the sequence are intended to cover more advanced and specialized topics in equilibrium and nonequilibrium statistical mechanics.

Any book, especially a slim one, on a subject as vast as thermodynamics and statistical physics, must be highly selective. In making this selection, I have attempted to adopt an approach that is largely model independent and to present a more detailed integration of classical thermodynamics and quantum theory than is attempted in comparable contemporary texts. The nature of such a book is determined mainly by the illustrations of fundamental theory that are chosen; and a book much like the present one could not have been written before the early sixties although the foundations of this subject were completed thirty years earlier.

In the first part of this book, classical thermodynamics is presented as a rigorous, closed discipline that is firmly and independently based on experiment. We believe that this traditional arrangement is logically the best, and physically the most natural, since it uses macroscopic concepts to discuss macroscopic phenomena. The second part of the book is devoted to a derivation of the macroscopic properties of large systems from atomic theory or, in other words, to the thermodynamic manifestations of quantum theory. Particular emphasis is placed on so-called macroscopic quantum systems such as superfluids.

The detailed development of the classical and quantal many-body formalism is regarded as belonging to the second course of the sequence and beyond the scope of this book; however, an introductory discussion of phase transitions according to statistical mechanics is given.

The detailed discussions of transport phenomena and the Boltzmann problem form the third course of the sequence; so, here again, only an introduction has been attempted.

Although no problem sets are included, the subject matter of the book offers many possibilities, both for additional development and as a source of specific problems. The author hopes that some will enjoy the flexibility of such a text.

I would like to thank many colleagues, particularly N. Byers, B. Fried, A. Glassgold, R. Norton, M. Revzen, M. Ruderman, and I. Rudnick, for their helpful comments on the manuscript.

September 1968
Los Angeles, California   ROBERT J. FINKELSTEIN

Thermodynamics and Statistical Physics

1

The Zeroth and the First Law

1.1 Introduction

The first law of thermodynamics was established about 1850, primarily by the theoretical considerations of Mayer and the experimental investigations of Joule. Since the essential content of the second law had been discovered earlier by Carnot, the formal development of thermodynamics proceeded rapidly, mainly in the work of Kelvin and Clausius, and was completed by 1878 in Gibbs' paper "On the Equilibrium of Heterogeneous Substances." The statistical foundations of thermodynamics, which had been first explored by Maxwell and Boltzmann, were essentially understood by 1900 in the work of Gibbs and Einstein—only slightly before the dramatic discoveries that launched atomic physics and realistically implemented the statistical formalism that had so recently been prepared. These developments mark the historical path connecting classical macroscopic physics with modern atomic physics.

One may enter on this path by recognizing that the classical equations of hydrodynamics are incomplete and that it is necessary to have some way of dealing with the intuitive concepts of temperature and heat in order to make predictions about common macroscopic processes. From the microscopic point of view, any description of the large collection of molecules constituting a fluid by only a few macroscopic parameters is extremely incomplete. Thermodynamics is made possible by the fact that there exists a large domain of physics where the influence of all these neglected microscopic variables may be approximated by adjoining the simple thermal concepts to classical mechanics.

It is logically most natural to define these thermal concepts in terms of the mechanical concepts that they are supplementing. This procedure, which is due to Born and Carathéodory, is the general approach to be followed here [1–3]. However, we shall not follow the extreme axiomatic formulation [2].

1.2   Temperature

There is an interesting analogy between the roles of time in mechanics and temperature in thermodynamics. Psychological time and temperature grow out of one-dimensional orderings, which are of physiological origin and are suggested by the words "after" and "warmer" respectively. With the aid of simple clocks and thermometers these subjective orderings were long ago made the basis of objective concepts, but the subsequent development of these concepts had to wait until quite recent times for the discovery of certain empirical laws, namely: the principle of relativity, and the second law of thermodynamics, respectively.

The concept of time did not emerge in its modern form until it was concluded on the basis of many experiments that absolute motion could not be detected in any way at all. Similarly, the concept of absolute temperature was not developed before it was discovered that certain kinds of thermal processes, which are described by the second law, were in principle unobservable. In order to define time and temperature in ways that are at present considered satisfactory one has to postulate the principle of relativity in the one case and the second law of thermodynamics in the other. Indeed the problems of defining time and temperature are essentially the same as the tasks of working out the formalisms of special relativity and thermodynamics respectively.

1.3   The Reduction of Thermal to Mechanical Concepts

We shall require, following Born [1], that all thermal concepts, including temperature, be defined in terms of mechanical variables only. To make a beginning in this direction, let A be a well-defined physical system, completely characterized by the set of mechanical variables $\mathbf{A}$. For example, if A is a simple homogeneous, isotropic fluid enclosed in a container, then $\mathbf{A} = (p_a, V_a, M_a)$ where (p_a, V_a, M_a) are pressure, volume, and mass. Systems of more complexity, for example, nonhomogeneous, anisotropic masses, will require more variables.

We define first a *steady state* by the condition that all variables are time independent.

Steady state:

$$\frac{\partial \mathbf{A}}{\partial t} = 0. \tag{3.1}$$

We next define *thermal contact* and *thermal isolation*. Consider two systems, A and B, both in steady states. Then there are two logical possibilities:
   (a) the variables $\mathbf{A}$ and $\mathbf{B}$ are independent or
   (b) they are codetermined.
In case (a) we say that A and B are *thermally isolated*. If A and B are also in geometrical contact through a wall, we say that they are separated by an insulating, or *adiabatic*, wall.
   In case (b) we say that the systems A and B are in *thermal contact* or *thermal equilibrium* with each other.

Thermal equilibrium:

$$F_{ab}(\mathbf{A}, \mathbf{B}) = 0 \tag{3.2}$$

where equation (3.2) expresses the fact that the variables $\mathbf{A}$ and $\mathbf{B}$ are codetermined and F_{ab} can depend only on the systems A and B.
   In case (*b*), if the systems are in contact through a wall, we speak of a conducting, or *diathermic*, wall. If they are not in geometrical contact, we may speak of thermal equilibrium at-a-distance, such as exists between distant spheres which are in radiative equilibrium in otherwise empty space.
   The adiabatic and the diathermic wall of course represent idealizations of insulating and conducting walls. An actual wall may be judged insulating for sufficiently short periods of contact and conducting for sufficiently long intervals [4].

1.4   Zero Law and Thermal Equation of State

The important empirical fact about thermal equilibrium is that this relation is transitive. Since it is symmetrical and reflexive as well, we shall use the equality sign to express thermal equilibrium or contact:

$$A = B. \tag{4.1}$$

The transitive property of thermal equilibrium is the zero law of thermodynamics.

Zero law:

$$\text{If } A = B, \text{ and } B = C, \text{ then } A = C. \tag{4.2}$$

This statement implies that any arbitrary system C that is chosen as a thermometer will read the same for A and B, if A and B are in thermal contact.

The zero law may also be stated in the notation of equation (3.2) as follows: If

$$F_{ab}(\mathbf{A}, \mathbf{B}) = 0 \tag{4.3a}$$

and

$$F_{bc}(\mathbf{B}, \mathbf{C}) = 0, \tag{4.3b}$$

then

$$F_{ac}(\mathbf{A}, \mathbf{C}) = 0. \tag{4.3c}$$

In general, equation (4.3c) is not a mathematical consequence of (4.3a) and (4.3b) for arbitrary functions. On the other hand the zero law asserts that (4.3c) always does follow. The zero law therefore restricts the functions F. In fact it is possible to eliminate B in general only if the various components of B are locked together in one function, say G. Then

$$F_{ab}(\mathbf{A}, G(\mathbf{B})) = 0 \tag{4.4a}$$

and F_{ab} may be formally inverted as follows:

$$G(\mathbf{B}) = H(\mathbf{A}) \tag{4.4b}$$

where (4.4a) and (4.4b) describe the same functional relation. Adopting a better notation, we now see that the relation (3.2) or (4.4b) must be of the form

$$f_a(\mathbf{A}) = f_b(\mathbf{B}). \tag{4.5}$$

This relation permits one to define temperature as follows. Either A or B may be considered the thermometer; suppose we choose B. Then by measuring the variables $\mathbf{B}$ one may find $f_b(\mathbf{B})$ and define $t_b(\mathbf{A}) \equiv f_b(\mathbf{B})$ to be the temperature of A as measured by the thermometer B. Then

$$f_a(\mathbf{A}) = t_b(\mathbf{A}) \tag{4.6}$$

may be called a *thermal equation of state* of A. If any two systems, A and C, have the same B-temperature, then they are in thermal equilibrium with each other. Furthermore, if A and C are in equilibrium according to a B thermometer they are also in equilibrium according to any other thermometer. Thus although the numerical value of the temperature depends on the thermometer, a statement of the equality of the temperatures, say,

$$t_b(\mathbf{A}) = t_b(\mathbf{C}) \tag{4.7}$$

remains true under a change of thermometer. Any temperature defined in the preceding way is called an *empirical temperature*.

Examples

When gases are in thermal equilibrium we have approximately, for equal numbers of mols,

$$P_a V_a = P_b V_b \tag{4.8a}$$

and in the next approximation,

$$f(P_a V_a \alpha_a \beta_a) = f(P_b V_b \alpha_b \beta_b) \tag{4.8b}$$

where α and β are Van der Waals constants. The gas temperature may be defined by

$$t = f(P, V, \alpha, \beta) \tag{4.9}$$

and will differ slightly depending on the α and β. An ideal gas is an idealization defined to have the thermal equation of state

$$PV = Rt \tag{4.10}$$

where R is a constant. Then t is the ideal gas temperature.

In general the set of all states of a system A that are in equilibrium with each other satisfy by (4.6):

$$f_a(\mathbf{A}) = t_a.$$

States belonging to a given value of t_a lie on the same hypersurface in the space of the mechanical variables $(\mathbf{A})$. The different hypersurfaces may be ordered by the value of t_a since they can not intersect. Similarly we have for another system B

$$f_b(\mathbf{B}) = t_b.$$

The transformation between the A-temperature, t_a, and the B-temperature, t_b, must be a one-to-one correspondence to avoid violation of the zeroth law. Therefore the transformation between one empirical temperature and another will preserve the temperature order. It will be seen in chapter 3 how the temperature order associated with the zeroth law is made more definite by the second law.

Finally one may speak of a state of thermal equilibrium of an *isolated* system; if such a system is put in thermal contact with any other system that has the same empirical temperature, all mechanical variables remain unchanged.

1.5   Quasi-static and Adiabatic Processes

We extend the idea of thermal equilibrium and thermal isolation from *state* to *process* as follows. A process carried out so slowly that it approximates a sequence of *equilibrium* states will be called *quasi-static*. Therefore any change that can be described by a curve on a *PV*-diagram (or more generally in the space of the complete set of mechanical variables) is quasi-static. On the other hand, any process, quasi-static or not, carried out on a *thermally isolated* system will be called *adiabatic*.

1.6   First Law and Caloric Equation of State

Whereas the zero law concerns systems in thermal contact, the first law takes its simplest form for a system in thermal isolation. One has the following empirical induction.

First Law:

The mechanical work done in an adiabatic process, i.e., in taking an *isolated* system from the initial state (A_1) to the final state (A_2), is independent of the manner in which this work is done and depends only on the states A_1 and A_2.

With the aid of the first law it is possible to define a new function $U(A)$:

$$U(A) = U(A_0) + W_{ad}(A, A_0). \tag{6.1}$$

This function $U(A)$, which is called the internal energy, is defined only up to the additive constant $U(A_0)$ where A_0 is an arbitrary standard state. Assigning $U(A_0)$, one may then experimentally define $U(A)$ for any other state A by measuring the mechanical work $W_{ad}(A, A_0)$ in an *adiabatic* (but not necessarily quasi-static) process.

The empirical function so determined is called the internal energy function, and

$$U = U(A) \tag{6.2}$$

is called the caloric equation of state. The thermal equation of state already given may be written

$$t = t(A). \tag{6.3}$$

These two empirical equations of state determine the thermodynamic properties of A. They have been defined entirely in terms of mechanical variables, according to the following logical scheme:

Thermal contact $\longrightarrow$ Zero law $\longrightarrow$ Thermal equation of state
Thermal isolation $\longrightarrow$ First Law $\longrightarrow$ Caloric equation of state.

QUANTITY OF HEAT

If the system is not enclosed by an adiabatic wall, it is found in general that the work done is no longer determined by the initial and final states alone, i.e.,

$$W_{12} \neq U_1 - U_2. \tag{6.4}$$

The energy discrepancy is defined to be the heat absorbed or given out in the change:

$$\Delta Q \equiv \Delta U + \Delta W \tag{6.5}$$

where ΔW means the work done by the system. Only when the change is adiabatic do we have $\Delta Q = 0$. Then

$$\Delta U = -\Delta W_{\text{adiabatic}}. \tag{6.6}$$

In general,

$$\Delta Q = \Delta W - \Delta W_{\text{adiabatic}}. \tag{6.7}$$

The preceding equation is the macroscopic definition of heat. The first law has a simple kinetic interpretation in terms of the conservation of energy: namely ΔQ is the total energy absorbed by the molecules, while ΔW is the macroscopic work done by the molecules.

1.7   Some Important Applications to Ideal Gas Systems

We illustrate the concepts already introduced with some remarks about ideal gases. These simple systems are of fundamental importance since all forms of matter approximate ideal gas behavior at high temperatures and/or low densities.

EQUATIONS OF STATE

An ideal gas is defined by the following thermal and caloric equations:

$$t \sim PV \tag{7.1}$$
$$U \sim PV. \tag{7.2}$$

Hence

$$U = Ct \tag{7.3}$$

where C is a constant. Let us introduce the usual notation for the constant in the thermal equation. Then

$$PV = Rt \qquad (7.4)$$

where $R = 2$ cal/mol deg.

The first law for this case is

$$dQ = dU + P \, dV = C \, dt + P \, dV.$$

Define the specific heat at constant volume:

$$C_v = \left(\frac{dQ}{dt}\right)_V. \qquad (7.5)$$

Then

$$C_v = C$$

and

$$dQ = C_v dt + P \, dV. \qquad (7.6)$$

Define the specific heat at constant pressure:

$$C_p = \left(\frac{dQ}{dt}\right)_P. \qquad (7.7)$$

Then

$$C_p = C_v + R \qquad (7.8)$$

and

$$dQ = C_p dt - V \, dP \qquad (7.9)$$

by the thermal equation of state.

ADIABATIC CHANGES

An adiabatic change is defined by the condition

$$dQ = 0.$$

For such a change:

$$C_p dt - V \, dP = 0$$
$$C_v dt + P \, dV = 0$$
$$\frac{V \, dP}{P \, dV} = -\frac{C_p}{C_v}.$$

If the total change takes place through a succession of equilibrium states, we may integrate. Then [5]

$$PV^\gamma = \text{constant} \tag{7.10}$$

where

$$\gamma = C_p/C_v. \tag{7.11}$$

If the change takes place so rapidly that a succession of intermediate equilibrium states is not set up, then the variables P and t are not defined for these states or may not be related according to equation (7.4) and the above integration can not be carried out. Therefore equation (7.10) holds only for a quasi-static adiabatic change.

POLYTROPIC CHANGES

If the specific heat remains constant, a quasi-static process is called *polytropic*. We have, if C is the specific heat,

$$C\,dt = C_v dt + P\,dV$$
$$C\,dt = C_p dt - V\,dP$$

or

$$\frac{V}{P}\frac{dP}{dV} = -\gamma' \tag{7.12}$$

where

$$\gamma' = \frac{C_p - C}{C_v - C}. \tag{7.13}$$

Therefore

$$PV^{\gamma'} = \text{const.,} \tag{7.14}$$

again if the total change takes place through a succession of intermediate equilibrium states.

In the adiabatic and isothermal cases we have $C = 0$, $\gamma' = \gamma$ and $C = \infty$, $\gamma' = 1$, respectively.

VELOCITY OF SOUND

According to the equations of hydrodynamics, i.e., according to the conservation of mass and momentum, the velocity of an infinitesimal pulse is

$$c = \left(\frac{dP}{d\rho}\right)^{1/2} \tag{7.15}$$

where ρ is the density. This formula does not lead to a definite result unless a particular relation between P and ρ is assumed. If $P \sim \rho^{\gamma'}$, then

$$c = \left(\frac{\gamma' P}{\rho}\right)^{1/2}. \tag{7.16}$$

Under polytropic conditions, γ' lies between 1 and γ depending on the frequency, but is usually quite close to γ.

POLYTROPIC CYCLE

Suppose we have a cycle composed of two pairs of polytropic lines. Let

$$\phi_\gamma(i) = P_i V_i^\gamma.$$

Then if the points are paired by the γ-curves, we have in Figure 1.1

$$\phi_\gamma(1) = \phi_\gamma(2)$$
$$\phi_\gamma(3) = \phi_\gamma(4)$$

and

$$\phi_\gamma(1) \, \phi_\gamma(3) = \phi_\gamma(2) \, \phi_\gamma(4).$$

Similarly, if the pairing is done with respect to the γ'-curves, we have

$$\phi_{\gamma'}(1) \, \phi_{\gamma'}(3) = \phi_{\gamma'}(2) \, \phi_{\gamma'}(4)$$

and therefore

$$\frac{\phi_\gamma(1) \, \phi_\gamma(3)}{\phi_{\gamma'}(1) \, \phi_{\gamma'}(3)} = \frac{\phi_\gamma(2) \, \phi_\gamma(4)}{\phi_{\gamma'}(2) \, \phi_{\gamma'}(4)}$$

or

$$\left(\frac{V_1 V_3}{V_2 V_4}\right)^{\gamma' - \gamma} = 1 \tag{7.17}$$

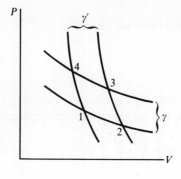

FIGURE 1.1
Polytropic cycle.

where $\gamma' \neq \gamma$. It then follows from equation (7.17) that

$$V_1 V_3 = V_2 V_4. \tag{7.18}$$

CARNOT CYCLE

The Carnot cycle is the special case in which one pair of lines is isothermal and the other adiabatic.

Let us consider a perfect gas enclosed in a cylinder and suppose that it is carried through a Carnot cycle in the clockwise direction so that the total work done by the gas is positive. We may compute its efficiency as follows:

$$\oint dQ = \oint dU + \oint dW$$
$$= \oint dW$$

or

$$Q_> + Q_< = W$$

where $Q_<$ and $Q_>$ are the heats absorbed at the lesser ($t_<$) and greater ($t_>$) temperatures, respectively. Hence the efficiency is

$$\eta = W/Q_> = 1 + Q_</Q_>.$$

Since the internal energy depends only on the temperature, it does not alter in the isothermal part of the cycle. Hence

$$Q_> = \int_{V_4}^{V_3} P\, dV = Rt_> \ln \frac{V_3}{V_4}$$
$$Q_< = \int_{V_2}^{V_1} P\, dV = Rt_< \ln \frac{V_1}{V_2}.$$

Hence

$$\eta = 1 + \frac{t_<}{t_>} \frac{\ln(V_1/V_2)}{\ln(V_3/V_4)} = 1 - \frac{t_<}{t_>}, \tag{7.19}$$

since

$$V_1 V_3 = V_2 V_4.$$

POLYTROPIC GAS SPHERE IN GRAVITATIONAL EQUILIBRIUM

Consider an isolated mass of gas. In terrestrial problems, such a system is usually contained by solid walls. Let us consider, however, a gaseous sphere: for example, a star that is prevented from further expansion by its own

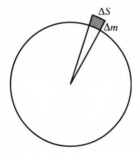

FIGURE 1.2
Gas sphere in gravitational equilibrium.

gravitational field. The equation of hydrostatic equilibrium for a small mass Δm, with thickness Δr, and surface ΔS, is (see Figure 1.2)

$$-\frac{dP}{dr}\,\Delta r\,\Delta S = \frac{GM}{r^2}\,\Delta m.$$

Here dP/dr is the pressure gradient, G is the gravitational constant, and M is the mass contained in the spherical surface $(4\pi r^2)$. Then $\Delta m = \rho\,\Delta S\Delta r$ where ρ is the density.

The equilibrium equations may therefore be written as

$$\frac{GM\rho}{r^2} = -\frac{dP}{dr} \qquad (7.20)$$

or

$$4\pi\,Gr^2\rho = -\frac{d}{dr}\left(\frac{r^2}{\rho}\frac{dP}{dr}\right). \qquad (7.21)$$

Equation (7.21) is a differential equation in two dependent variables, namely P and ρ. In order to formulate a determinate mathematical problem it is necessary to assume a relation between P and ρ. One gets a simple model of a star by assuming a polytropic connection between P and ρ:

$$P \sim \rho^\gamma. \qquad (7.22)$$

After such a relation is assumed, it is then possible to find the density distribution provided that the boundary conditions on the differential equation are given. These boundary conditions may be chosen to be, say, the central density or central temperature. By symmetry

$$\left(\frac{dP}{dr}\right)_{r=0} = 0. \qquad (7.23)$$

Then one may find the density distribution:

$$\rho = \rho(t_0, r) \qquad (7.24)$$

where t_0 is the central temperature. One may also find

$$M = M(t_0) \tag{7.25a}$$

$$R = R(t_0) \tag{7.25b}$$

where M is the total mass of the star and R is some suitably defined radius. In this way one gets a one-parameter (t_0) family of solutions and if t_0 is eliminated, one may also find the relation between mass and radius for the assumed model:

$$M = M(R). \tag{7.26}$$

Stellar models of this kind are discussed in detail by Chandrasekhar [3].

If actual stars were described by such a simple model, then they would lie on a single curve in the mass-luminosity (Hertzsprung-Russell) diagram. Such a model neglects radiation pressure, internal energy sources and also heterogeneous chemical composition, and is therefore unrealistic in these respects. The main sequence stars lie on a band instead of a curve.

1.8 Completion of Hydrodynamical Equations by Thermodynamics

Examples like the preceding model of a star and equation (7.15) for the velocity of sound reveal that the hydrodynamic equations are incomplete. Although one may arrive at a mathematically determinate problem in both of these cases by assuming a relation like (7.22), there is in the discussion so far no basis for such an assumption. We shall now see how our equations may be completed in a physically justifiable way.

The usual hydrodynamic equations are

$$\frac{\partial \rho}{\partial \tau} + \text{div} (\rho \mathbf{u}) = 0 \tag{8.1}$$

$$-\text{grad } P = \rho \frac{D\mathbf{u}}{D\tau} \tag{8.2}$$

where

$$\frac{D}{D\tau} = \frac{\partial}{\partial \tau} + \mathbf{u} \text{ grad} \tag{8.3}$$

is the hydrodyanmical derivative, and τ is the time.

Equation (8.1) is the equation of continuity and (8.2) is the equation of motion. There are only four differential equations for the five unknown functions $(P, \rho, \mathbf{u})$. Therefore the hydrodynamical equations are incomplete.

The one missing differential equation is provided by the first law of thermo-dynamics:

$$\frac{DQ}{D\tau} = \frac{DU}{D\tau} + P\frac{DV}{D\tau}.$$ (8.4)

In order not to introduce at the same time additional unknown functions, we must assume knowledge of the caloric and thermal equations of state:

$$U = U(P, V)$$ (8.5)

$$t = t(P, V).$$ (8.6)

Here V means the specific volume and is related to the density as follows:

$$\rho V = 1.$$ (8.7)

If chemical reactions are taking place in the fluid, then $DQ/D\tau$ depends on the thermal equation of state. The situation may be further complicated if thermal conduction and viscosity are important.

It is of course easy to find situations in which the above equations are also incomplete. For example, the motions may become sufficiently violent so that equilibrium conditions do not exist and the equations of state are not satisfied. The basic point is that the complete molecular system has many more degrees of freedom than the approximating macroscopic system and will violate the hydrodynamical equations if these degrees of freedom are excited.

1.9   Application to Simple Fluid Motion
(One-dimensional, Steady, and Locally Adiabatic)

In order to avoid geometrical complication consider a one-dimensional steady flow which is also locally adiabatic. If the flow is one-dimensional, one has

$$\frac{\partial\rho}{\partial\tau} + \frac{\partial}{\partial x}(\rho u) = 0$$ (9.1)

$$-\frac{\partial P}{\partial x} = \rho\frac{Du}{D\tau}$$ (9.2)

$$\frac{DQ}{D\tau} = \frac{DU}{D\tau} + P\frac{DV}{D\tau}$$ (9.3)

where

$$\frac{D}{D\tau} = \frac{\partial}{\partial\tau} + u\frac{\partial}{\partial x}.$$

Let us consider a very simple situation, namely, steady states which are locally adiabatic, i.e.,

$$\frac{\partial \rho}{\partial \tau} = \frac{\partial u}{\partial \tau} = 0 \tag{9.4}$$

and

$$\frac{DQ}{D\tau} = 0. \tag{9.5}$$

Then the hydrodynamic equations become

$$\frac{\partial}{\partial x}(\rho u) = 0 \tag{9.6}$$

$$\frac{\partial P}{\partial x} + \rho u \frac{\partial u}{\partial x} = 0 \tag{9.7}$$

with the following integrals:

$$\rho u = \text{constant} \tag{9.8}$$

$$P + \rho u^2 = \text{constant}. \tag{9.9}$$

The energy equation may be written, by (9.5),

$$0 = \frac{\partial U}{\partial x} + P \frac{\partial V}{\partial x}. \tag{9.10}$$

Let us now combine equations (9.10) and (9.2) by first re-expressing (9.2) as follows:

$$V \frac{\partial P}{\partial x} + \frac{\partial}{\partial x}\left(\frac{u^2}{2}\right) = 0. \tag{9.11}$$

Adding (9.11) and (9.10) one gets

$$\frac{\partial}{\partial x}\left(U + PV + \frac{u^2}{2}\right) = 0 \tag{9.12}$$

with the integral

$$U + PV + \frac{u^2}{2} = \text{constant}. \tag{9.13}$$

The three integrals (9.8), (9.9), and (9.13) express conservation of mass, momentum, and energy. The conservation of energy condition expresses clearly the coupling of the macroscopic fluid motion to the internal microscopic motion.

If we had not considered such an idealized situation, none of the essentials

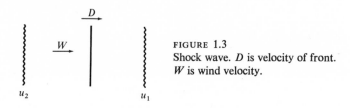

FIGURE 1.3
Shock wave. D is velocity of front.
W is wind velocity.

would have been changed but the mathematical analysis would have become more complicated [6]. Even though these equations are very idealized, we shall now make a practical application of them.

1.10   Shock Waves

A shock wave is a propagating mechanical and thermal discontinuity. Shocks are sometimes produced by explosions and the motion of supersonic solids; however any disturbance at all tends to evolve into a shock because the high pressure regions tend to move faster and overtake the front.

Let us consider a steady shock front moving at constant velocity; for example, such a shock is carried by a supersonic airplane. Let D be its velocity, let W be the wind velocity behind the front (Figure 1.3). By steady we mean that the thermal and mechanical variables are time independent when described in a coordinate frame attached to the front. Since this front is not accelerated, the hydrodynamic equations are unchanged in this coordinate system. Under the conditions just stated we have the following relations by (9.8), (9.9), and (9.13):

$$\rho_1 u_1 = \rho_2 u_2 \tag{10.1}$$

$$P_1 + \rho_1 u_1{}^2 = P_2 + \rho_2 u_2{}^2 \tag{10.2}$$

$$U_1 + P_1 V_1 + \frac{u_1{}^2}{2} = U_2 + P_2 V_2 + \frac{u_2{}^2}{2} \tag{10.3}$$

where 1 and 2 refer to the two sides of the shock front. Here u_1 and u_2 represent the mass velocities on the two sides of the front referred to the comoving coordinate system. Hence

$$u_2 = D - W \tag{10.4}$$

$$u_1 = D. \tag{10.5}$$

We shall see that at the shock front there is a discontinuity in each of the mechanical and thermal variables.

By combining the first two conditions (conservation of mass and momentum) we obtain

$$P_2 - P_1 = \rho_1 u_1{}^2 - \rho_2 u_2{}^2$$

$$= \rho_1 u_1{}^2 - \rho_2 \left(\frac{\rho_1 u_1}{\rho_2}\right)^2$$

$$u_1 = V_1 \left(\frac{P_2 - P_1}{V_1 - V_2}\right)^{1/2} \tag{10.6}$$

$$u_2 = V_2 \left(\frac{P_2 - P_1}{V_1 - V_2}\right)^{1/2}. \tag{10.7}$$

These equations give the shock and wind velocities. In particular

$$D = V_1 \left(\frac{P_2 - P_1}{V_1 - V_2}\right)^{1/2}. \tag{10.8}$$

In the limiting case of a weak shock we have

$$\lim_{P_2 \to P_1} D = V \left(-\frac{dP}{dV}\right)^{1/2} = \left(\frac{dP}{d\rho}\right)^{1/2} \tag{10.9}$$

which is the usual formula for the velocity of sound. The energy condition can be simplified if the kinetic energy terms are eliminated:

$$U_2 - U_1 = \frac{1}{2} u_1{}^2 - \frac{1}{2} u_2{}^2 + P_1 V_1 - P_2 V_2$$

$$= \frac{1}{2} (V_1{}^2 - V_2{}^2) \left(\frac{P_2 - P_1}{V_1 - V_2}\right) + P_1 V_1 - P_2 V_2$$

$$U_2 - U_1 = \frac{1}{2} (P_1 + P_2)(V_1 - V_2). \tag{10.10}$$

In the limit of weak shocks this becomes

$$dU + P \, dV = 0. \tag{10.11}$$

For the finite case we have, however, equation (10.10). This relation is called the Rankine-Hugoniot equation.

If terms representing thermal conductivity and viscosity had been carried in the fundamental differential equations, then even with the other simplifying assumptions that have been made in the preceding paragraphs the final equations (10.1) (10.3) would have been differential rather than algebraic relations. By solving these differential equations one may find interpolating functions which connect the hydrodynamic variables on the two sides of the shock and at the same time it is possible to make a crude estimate of the thickness of the shock zone [7].

SHOCKS IN IDEAL GASES

It is convenient to introduce the dimensionless variables:

$$x = \frac{P_2}{P_1}, \qquad y = \frac{V_2}{V_1}, \qquad z = \frac{t_2}{t_1}. \tag{10.12}$$

Then (10.8) becomes

$$D = (P_1 V_1)^{1/2} \left(\frac{x-1}{1-y} \right)^{1/2}.$$

For an ideal gas at pressure P_1 and specific volume V_1, the velocity of sound is

$$c = (\gamma P_1 V_1)^{1/2}. \tag{10.13}$$

Hence

$$M = \frac{D}{c} = \frac{1}{\sqrt{\gamma}} \left(\frac{x-1}{1-y} \right)^{1/2}. \tag{10.14}$$

M is called the Mach number.

Let us rewrite the Rankine-Hugoniot equation as well:

$$f_2 xy - f_1 = (1 + x)(1 - y)$$

where

$$f = \frac{2}{\gamma - 1}$$

is the effective number of degrees of freedom of each molecule [5]. Then

$$y = \frac{1 + f_1 + x}{1 + (1 + f_2)x}. \tag{10.15}$$

From equations (10.14) and (10.15) we may for example compute M and y in terms of x. The temperature ratio across the front is

$$z = xy. \tag{10.16}$$

1.11   Characteristic Functions

The possibility of analyzing a shock in the way just described depends upon the fact that the change in the internal energy function depends only on the initial and final states and not on the details of the process connecting them. This, in fact, is the content of the first law under adiabatic boundary conditions, such as we have here assumed. Functions that depend only on the state and not on the way the state is set up are obviously of great importance; they are called characteristic functions and so far we have found two examples: the temperature and the internal energy. In contrast, the heat and the work are not. The two characteristic functions, temperature and internal energy, exist by virtue of the zero and first laws, respectively. We shall now begin the discussion of the second law of thermodynamics, which permits us to define a third characteristic function, the entropy.

Notes and References

1. Max Born, *Natural Philosophy of Cause and Chance* (Oxford Univ. Press, London, 1948).
2. P. T. Landsberg, *Thermodynamics* (Interscience, London, 1961).
3. S. Chandrasekhar, *An Introduction to the Theory of Stellar Structure* (University of Chicago Press, Chicago, 1939). The Born-Carathéodory approach is discussed in references [1–3].
4. Thermal contact and thermal equilibrium are used synonomously here if the system has reached a steady state. If the system has *not* yet reached a steady state, we speak of thermal contact.
5. The numerical value of γ is severely restricted by kinetic theory. According to the equipartition theorem,

$$U = N\tfrac{1}{2}(fkT),$$

where f is the number of degrees of freedom of a single molecule and N is the number of molecules in the gas. Then

$$C_v = (\tfrac{1}{2})Nkf = \tfrac{1}{2}Rf.$$

Hence

$$\gamma = C_p/C_v = 1 + R/C_v = 1 + 2/f.$$

For monatomic molecules, $f = 3$ and $\gamma = 1.67$.
For very large molecules, $f \longrightarrow \infty$ and $\gamma \longrightarrow 1$.
6. See, for example, R. Courant and K. O. Friedrichs, *Supersonic Flow and Shock Waves* (Interscience, New York, 1948).
7. See, for example, *Modern Developments in Fluid Dynamics*, High Speed Flow, Vol. I, p. 122 (Oxford Univ. Press, London, 1953).

2

The Second Law

2.1 The Second Law of Thermodynamics

We have seen how temperature and heat may be defined in terms of mechanical concepts with the aid of the zero and first laws. The equations of hydrodynamics may then be completed with the aid of the first law and the thermal and caloric equations of state. After this is done, one obtains a set of equations like (10.1)-(10.3) of Chapter 1, which express the conservation of mass, momentum, and energy. These equations, however, do not even touch upon the really central feature of thermodynamics which is expressed by the second law and two new characteristic functions, *entropy* and *absolute temperature*.

The second law has received many formulations that illuminate its different facets and give different kinds of insight into a most comprehensive induction from experience. The first formulation we shall discuss emphasizes its mathematical structure, and is the following:

The imperfect differential dQ, constructed for an infinitesimal quasi-static change, possesses an integrating factor.

This statement must be understood in the following way. Consider two nearby states of thermal equilibrium. Then by definition of dQ, we have

$$dQ = dU + dW$$
$$= \sum_i X_i \, dx_i \tag{1.1}$$

where x_i is the complete set of variables needed to specify the system and X_i are the appropriate functions needed to construct dU and dW. The second law now tells us that this differential form has an integrating factor.

The integrating denominator is defined to be the absolute temperature, T, and the perfect differential dQ/T is called the entropy.

A differential form in two variables always has an integrating denominator. In such simple systems the existence of an integrating denominator is mathematically guaranteed and therefore is not a physical constraint. We shall illustrate these remarks by first considering a perfect gas.

THE ENTROPY OF AN IDEAL GAS

We have already seen that the following relation holds for an ideal gas operating in a Carnot cycle:

$$\frac{Q_1}{t_1} + \frac{Q_2}{t_2} = 0. \tag{1.2}$$

This equation can also be expressed in the form

$$\oint \frac{dQ}{t} = 0 \tag{1.3}$$

where the integral is taken around the Carnot cycle. By paving an arbitrary closed curve in the pV-plane with Carnot cycles, we may show that (1.3) also holds for it. Hence we may say that

$$dS = \frac{dQ}{t}$$

is a perfect differential. The same remark follows if we simply divide the first law by t, i.e.,

$$dQ = C_V \, dt + P \, dV$$

$$\frac{dQ}{t} = C_V \frac{dt}{t} + P \frac{dV}{t}$$

$$= C_V \frac{dt}{t} + R \frac{dV}{V}$$

$$S = \int \frac{dQ}{t} = C_V \ln t + R \ln V.$$

We say that dQ is not a perfect differential but that it has an integrating factor, namely t^{-1}. However, there is no physics in this statement, because any differential form in two variables has an integrating factor.

2.2 Integrating Factor for Differential Form

We shall now consider a differential form in n variables in order to determine the condition that an integrating factor exists. Let

$$dQ = \sum_i X_i \, dx_i. \tag{2.1}$$

Such a form is called a Pfaffian. If this is a perfect differential, then

$$X_i = \frac{\partial Q}{\partial x_i}$$

and

$$\frac{\partial^2 Q}{\partial x_i \, \partial x_j} = \frac{\partial X_i}{\partial x_j} = \frac{\partial X_j}{\partial x_i}.$$

The condition that dQ is an exact differential may also be stated as follows:

$$\text{curl } X = 0. \tag{2.2}$$

By the n-dimensional Stokes theorem this means

$$\oint X \, dx = \oint \sum_i X_i \, dx_i = 0.$$

Let us next assume that dQ is not a perfect differential, but that it may be transformed into one by an integrating factor; i.e., we assume

$$\text{curl } X \neq 0$$

but that a λ exists such that

$$\text{curl } \lambda X = 0. \tag{2.3}$$

Then

$$\partial_i X_j - \partial_j X_i = (X_i \partial_j - X_j \partial_i)\phi \tag{2.4}$$

where

$$\phi = \ln \lambda.$$

The single function $\phi(x_1, \ldots, x_n)$ must satisfy the $\binom{n}{2}$ partial differential equations (2.4). In order that these equations be mutually consistent, certain compatibility equations must be satisfied. Multiply (2.4) by X_k and form the two other relations obtained by cyclic permutation:

$$X_k(\partial_i X_j - \partial_j X_i) = X_k(X_i \partial_j - X_j \partial_i)\phi$$
$$X_i(\partial_j X_k - \partial_k X_j) = X_i(X_j \partial_k - X_k \partial_j)\phi$$
$$X_j(\partial_k X_i - \partial_i X_k) = X_j(X_k \partial_i - X_i \partial_k)\phi. \tag{2.5}$$

By adding the equations (2.5) one finds the $\binom{n}{3}$ relations which do not contain the unknown function $\lambda(x_1, \ldots, x_n)$ and must be satisfied by the X_k among themselves:

$$\sum_{ijk} X_k(\partial_i X_j - \partial_j X_i) = 0. \tag{2.6}$$

In the two dimensional case the relation is satisfied identically since there is only one relation (2.4); hence there is always an integrating denominator. In the three dimensional case we have [1]

$$\mathbf{X} \operatorname{curl} \mathbf{X} = 0. \tag{2.7}$$

2.3　System with Three Degrees of Freedom

We have seen that a Pfaffian in two variables always has an integrating factor. We now consider a system with three degrees of freedom, for example, two bodies in thermal contact.

Suppose that each has two independent variables, say (P_a, V_a) and (P_b, V_b); the composite system has the variables $(P_a V_a; P_b V_b)$ but these are connected by the relation expressing thermal equilibrium:

$$f_a(P_a, V_a) = f_b(P_b, V_b). \tag{3.1}$$

Hence the composite system has three degrees of freedom, say V_a, V_b and $t = t_a = t_b$.

Consider a small displacement to another equilibrium state. Then

$$dQ = dQ_a + dQ_b$$
$$= dU_a + P_a \, dV_a + dU_b + P_b \, dV_b \tag{3.2}$$
$$= \left[\left(\frac{\partial U}{\partial V}\right)_a + P_a\right] dV_a + \left[\left(\frac{\partial U}{\partial V}\right)_b + P_b\right] dV_b + \left[\frac{\partial U_a}{\partial t} + \frac{\partial U_b}{\partial t}\right] dt.$$

We now have a Pfaffian in 3 variables; on mathematical grounds there may or may not be an integrating factor. It turns out however that there always is; but its existence is accounted for by a physical law, not by a mathematical theorem.

We shall now investigate the formal consequences of the existence of such a factor. We have

$$X_1 = \frac{\partial U_a}{\partial V_a} + P_a \qquad\qquad x_1 = V_a \qquad\qquad (3.3a)$$

$$X_2 = \frac{\partial U_b}{\partial V_b} + P_b \qquad\qquad x_2 = V_b \qquad\qquad (3.3b)$$

$$X_0 = \frac{\partial}{\partial t}(U_a + U_b) \qquad\qquad x_0 = t. \qquad\qquad (3.3c)$$

Let

$$[i, j] = \frac{\partial X_j}{\partial x_i} - \frac{\partial X_i}{\partial x_j}, \qquad\qquad (3.4)$$

then

$$[1, 2] = 0 \qquad\qquad (3.5a)$$

$$[2, 0] = -\frac{\partial P_b}{\partial t} \qquad\qquad (3.5b)$$

$$[1, 0] = -\frac{\partial P_a}{\partial t}. \qquad\qquad (3.5c)$$

By equation (2.7),

$$-X_1 \frac{\partial P_b}{\partial t} + X_2 \frac{\partial P_a}{\partial t} = 0 \qquad\qquad (3.6)$$

or

$$\left[\frac{1}{X}\frac{\partial P}{\partial t}\right]_b = \left[\frac{1}{X}\frac{\partial P}{\partial t}\right]_a = f(V_a, V_b, t). \qquad\qquad (3.7)$$

The first bracket is independent of V_a; the second bracket is independent of V_b; hence they are both independent of both V_a and V_b, and $f(V_a, V_b, t)$ must depend on t alone. Then we may write

$$f(t) = \left[\frac{(\partial P/\partial t)_V}{(\partial U/\partial V)_t + P}\right]_a = \left[\frac{(\partial P/\partial t)_V}{(\partial U/\partial V)_t + P}\right]_b. \qquad\qquad (3.8)$$

Here $(\partial P/\partial t)_V$ and $(\partial U/\partial V)_t + P$ each separately depend on the physical system, but the combination appearing in the bracket does not. This result has the following meaning:

Consider two nearby states of thermal equilibrium, say t and $t + \Delta t$ as measured by an arbitrary thermometer. Determine empirically $(\Delta P)_V$ for these states as well as $(\partial U/\partial V)_t$ and P. Here $(\Delta P)_V$ means that the two states being compared have the same volume. Then $\dfrac{(\Delta P)_V}{(\partial U/\partial V)_t + P}$ is independent of the physical system and of the t appearing in the denominator and defines a new measure of temperature difference—an absolute temperature in the sense that its numerical value is independent of the physical constitution of the system, as well as of the initial choice of empirical temperature.

The integrating factor itself satisfies the equations (2.4):

$$\left(X_1 \frac{\partial}{\partial x_2} - X_2 \frac{\partial}{\partial x_1}\right) \ln \lambda = 0 \tag{3.9a}$$

$$\left(X_2 \frac{\partial}{\partial x_0} - X_0 \frac{\partial}{\partial x_2}\right) \ln \lambda = -\left(\frac{\partial P_2}{\partial t}\right) \tag{3.9b}$$

$$\left(X_0 \frac{\partial}{\partial x_1} - X_1 \frac{\partial}{\partial x_0}\right) \ln \lambda = +\left(\frac{\partial P_1}{\partial t}\right). \tag{3.9c}$$

We may satisfy these equations by choosing

$$\ln \lambda = - \int f(t)\, dt. \tag{3.10}$$

Then λ is an integrating factor, $\lambda^{-1} = T$ is an integrating denominator, and dQ/T is a perfect differential. We then have

$$\frac{dT}{T} = f(t)\, dt$$

$$= \frac{(\partial P/\partial t)_V\, dt}{(\partial U/\partial V)_t + P}. \tag{3.11}$$

In order to complete the definition of T it is necessary to fix the single constant of integration. This may be done by assigning T^* arbitrarily for some definite physical state such as the triple point of water. Then the right hand side of the following equation,

$$\ln \frac{T}{T^*} = \int_{t^*}^{t} \frac{(\partial P/\partial t)_V\, dt}{(\partial U/\partial V)_t + P} = I(t, t^*), \tag{3.12}$$

is given empirically and T may be computed for any other physical state. Note that $I(t, t^*) = I(t', t'^*)$ where t' is any other empirical temperature. The temperature so defined is called the *absolute* temperature since the integral depends neither on the choice of t nor on the choice of physical system. The existence of such a function depends upon the physical fact that an integrating denominator does exist for dQ.

The condition for the existence of an integrating denominator when expressed in terms of T itself is by (3.11):

$$\left(\frac{\partial U}{\partial V}\right)_T + P = T\left(\frac{\partial P}{\partial T}\right)_V. \tag{3.13}$$

The left side of this equation is determined by the caloric equation of state;

the right side is determined by the thermal equation. The existence of an integrating denominator for dQ implies that the thermal and caloric equations are not independent.

2.4   The General Case

The preceding work has been carried through for two systems, each of which has two degrees of freedom. To discuss the general case let $x_0 = t$, and let $x_{i\alpha} = (V_i, \ldots)$ for $i = 1, \ldots, n$, where $x_{i\alpha}$ describes the complete set of variables needed to describe the ith subsystem. Then

$$dQ = dU + \sum_{i,\alpha} A_{i\alpha} \, dx_{i\alpha} \tag{4.1}$$

where the $A_{i\alpha}$ represent the generalized forces and the sum represents the total work. Then

$$dQ = X_0 \, dx_0 + \sum_{i=1}^{n} \sum_{\alpha} X_{i\alpha} \, dx_{i\alpha} \tag{4.2}$$

where

$$X_0 = \sum_{i=1}^{n} \left(\frac{\partial U_i}{\partial t} \right) \tag{4.2a}$$

$$X_{i\alpha} = \frac{\partial U_i}{\partial x_{i\alpha}} + A_{i\alpha} \tag{4.2b}$$

and

$$U_i = U_i(t, x_{i\alpha}) \tag{4.3}$$

$$A_{i\alpha} = A_{i\alpha}(t, x_{i\alpha}). \tag{4.4}$$

To apply the conditions (2.6) for the existence of an integrating denominator we again calculate

$$\frac{\partial X_{j\beta}}{\partial x_{i\alpha}} - \frac{\partial X_{i\alpha}}{\partial x_{j\beta}} = [i\alpha, j\beta]. \tag{4.5}$$

Clearly,

$$[i\alpha, j\beta] = 0 \quad \text{if} \quad i \neq 0, j \neq 0, i \neq j, \tag{4.6}$$

$$[i\alpha, i\beta] = \frac{\partial A_{i\beta}}{\partial x_{i\alpha}} - \frac{\partial A_{i\alpha}}{\partial x_{i\beta}} \tag{4.6a}$$

and

$$[0, i\alpha] = \frac{\partial X_{i\alpha}}{\partial t} - \frac{\partial}{\partial x_{i\alpha}} \left[\sum_{i=1}^{n} \left(\frac{\partial U_i}{\partial t} \right) \right]$$

$$= \frac{\partial A_{i\alpha}}{\partial t}. \tag{4.7}$$

The conditions (2.6) are empty unless one of the variables is t or unless $i = j$. In the first case

$$X_0[i\alpha, j\beta] + X_{i\alpha}[j\beta, 0] + X_{j\beta}[0, i\alpha] = 0$$

or

$$\frac{1}{X_{i\alpha}} \frac{\partial A_{i\alpha}}{\partial t} = \frac{1}{X_{j\beta}} \frac{\partial A_{j\beta}}{\partial t} = f(t). \tag{4.8}$$

This relation is the generalization of (3.8), and the argument proceeds as before. The general case therefore factors into a set of two-system equilibria and leads to no additional results. Although all representations of $f(t)$ are equivalent, it is natural to choose the simplest, namely (3.8).

The remaining equations, holding when $i = j$, are

$$\frac{\partial A_{i\alpha}}{\partial x_{i\beta}} = \frac{\partial A_{i\beta}}{\partial x_{i\alpha}}. \tag{4.9}$$

These are reciprocity relations connecting generalized forces and displacements.

2.5   Entropy

The preceding analysis permits one to define not only an absolute temperature (T) but also another characteristic function, the entropy, as follows:

$$dS = \frac{dQ}{T}. \tag{5.1}$$

Since T is an integrating denominator, dS is an exact differential. The first law written in terms of T and S reads

$$T \, dS = T_a \, dS_a + T_b \, dS_b + \cdots.$$

However

$$T_a = T_b = \cdots = T.$$

Therefore

$$dS = dS_a + dS_b + \cdots. \tag{5.2}$$

In discussing a composite system, $A + B$, where the parts have different temperatures one may write

$$\Delta Q_{A+B} = \Delta Q_A + \Delta Q_B.$$

But if $T_A \neq T_B$, it is no longer possible to divide this equation through by a common temperature. Nevertheless we still define

$$\Delta S_{A+B} = \Delta S_A + \Delta S_B.$$

We therefore define the entropy of a composite system as follows:

$$S_{A+B} = S_A + S_B \tag{5.3}$$

even when the separate parts are not in thermal equilibrium.

2.6 The Absolute Temperature and the Ideal Gas Temperature

We have

$$\ln \frac{T_2}{T_1} = \int_{t_1}^{t_2} \frac{(\partial P/\partial t)_V \, dt}{(\partial U/\partial V)_t + P}. \tag{6.1}$$

The right hand side may in principle be determined for *any* physical system since it is the same for all. It is only necessary to determine empirically the coefficients $(\partial P/\partial t)_V$ and $(\partial U/\partial V)_t$.

As an example of how one may proceed in principle, consider the ideal gas. Then

$$PV = Rt$$

$$U = C_V t$$

where t is the gas temperature, and

$$(\partial P/\partial t)_V = \frac{R}{V}$$

$$(\partial U/\partial V)_t = 0$$

$$\ln \frac{T_2}{T_1} = \int_{t_1}^{t_2} \frac{R}{V_P} \, dt = \ln \frac{t_2}{t_1}. \tag{6.2}$$

Hence the gas temperature t is always the same as the absolute temperature T if they agree for one state ($t_1 = T_1$). Since the ideal gas equations provide a fair description of matter under certain conditions, the ideal gas temperature gives an approximate realization of the absolute temperature.

2.7   Completion of Second Law

We have seen that the second law permits us to define two new functions: the absolute temperature and the entropy. Both are defined in terms of quasi-static processes, i.e., successions of equilibrium states. This means in particular that the whole process leading from the initial to the final state must have a macroscopic description; if molecular processes intervene in a way that cannot be described thermodynamically, then the change cannot be regarded as quasi-static. Thus the entropy is defined in terms of a particular kind of process, namely quasi-static, just as the internal energy is defined in terms of an adiabatic process. We now ask what form the second law takes if the process is not quasi-static, just as we asked earlier what form the first law takes when conditions are not adiabatic. The question may be answered by a similar procedure.

By a series of quasi-static experiments leading to a variable final state (f), we may determine the entropy function over a region of thermodynamic space according to the formula

$$S_f - S_0 = \underset{\text{qs}}{\int_0^f} \frac{dQ}{T} \qquad \text{quasi-static.} \qquad (7.1a)$$

If we then go from the initial state to f by some process that is not quasi-static, we find empirically

$$S_f - S_0 > \underset{\text{nqs}}{\int_0^f} \frac{dQ}{T} \qquad \text{nonquasi-static} \qquad (7.1b)$$

where T means the temperature of a large heat bath with which the system (itself not in thermal equilibrium) is in thermal contact and from which all the dQ flows.

It is also possible to follow a procedure like that used earlier to define dQ. We may write

$$\Delta S = \int \frac{dQ}{T} + \Delta S_i \qquad (7.2)$$

as the analogue of

$$\Delta U = -\Delta W + \Delta Q.$$

Here ΔS_i corresponds to ΔQ and is defined to be the discrepancy between $\int dQ/T$ and ΔS just as ΔQ is the difference between the work and the internal energy. Then experiment shows that

$$\Delta S_i = 0 \qquad \text{for quasi-static processes} \qquad (7.2a)$$

$$\Delta S_i > 0 \qquad \text{for nonquasi-static processes.} \qquad (7.2b)$$

Equation (7.2) may be regarded as a general formulation of the second law.

If the process is quasi-static, then $\Delta S_i = 0$; it is in addition possible to calculate $\int dQ/T$, and it is therefore also possible to calculate ΔS according to (7.2). For processes that are not quasi-static T refers to the temperature of the heat bath; for such processes it is not possible to calculate $\int dQ/T$ from the thermal and caloric equations of the system, since the system is not in thermal equilibrium with the bath. It is, however, still possible to determine ΔS for processes that are not quasi-static by constructing a quasi-static path between the same endpoints. We then have

$$\Delta S = \int_{qs} \frac{dQ}{T} = \int_{nqs} \frac{dQ}{T} + \Delta S_i \qquad (7.3)$$

where $\int_{qs}$ and $\int_{nqs}$ are the integrals of dQ/T for quasi-static and nonquasi-static processes joining the same endpoints. We shall call ΔS_i the irreversible increase in entropy, for reasons to be given in Section 2.9.

For processes that are adiabatic and nonquasi-static we have

$$\Delta S = \Delta S_i > 0. \qquad (7.4)$$

The total increase in entropy in an adiabatic process is therefore irreversible.

2.8 Examples of Entropy Changes in Nonquasi-static Processes

FREE EXPANSION

Consider a perfect gas which goes from (P_1V_1) to (P_2V_2) without doing any work or absorbing any heat (by diffusing through a porous wall). We may calculate the entropy change for this process as for any other (quasi-static or not) by simply comparing the entropies of initial and final states. We find

$$\Delta S = C_V \ln \frac{t_2}{t_1} + R \ln \frac{V_2}{V_1}. \qquad (8.1)$$

But

$$\Delta Q = \Delta U + \Delta W$$

and

$$\Delta Q = 0$$
$$\Delta W = 0$$

under the given conditions. Therefore by the preceding equation

$$\Delta U = 0$$

and by the caloric equation

$$\Delta t = 0.$$

Hence

$$\Delta S = R \ln \frac{V_2}{V_1}. \tag{8.2}$$

The empirical fact is that

$$V_2 > V_1.$$

Hence

$$\Delta S > 0. \tag{8.3}$$

This change cannot be described by a path in the PV-plane; for the change is accomplished microscopically and cannot be followed by keeping track of P and V only. Since the intermediate states therefore cannot be described macroscopically, they are not equilibrium states and the process can not be classified as quasi-static. This conclusion is in agreement with the second law, according to which the entropy increases in an adiabatic process that is not quasi-static.

THERMAL EQUALIZATION

Consider two identical perfect gases, A and B, at temperatures T_A and T_B. Assume they are brought together and reach thermal equilibrium at temperature T without doing any work. If the composite system, $A + B$, is isolated we have

$$\Delta Q_A + \Delta Q_B = 0$$

or

$$C(T - T_A) + C(T - T_B) = 0$$

where C is the specific heat. Then

$$T = \frac{1}{2}(T_A + T_B). \tag{8.4}$$

By the definition (5.3) of the entropy of a composite system

$$\Delta S = \Delta S_A + \Delta S_B$$

$$= \int_A \frac{dQ}{T} + \int_B \frac{dQ}{T}$$

$$= c_V \left[\ln \frac{T}{T_A} + \ln \frac{T}{T_B} \right]$$

$$= c_V \ln \frac{T^2}{T_A T_B}. \tag{8.5}$$

But

$$T^2 > T_A T_B$$

when (8.4) holds, and therefore

$$\Delta S > 0, \tag{8.6}$$

since empirically $c_V > 0$.

Thermal equalization is accomplished by molecular processes; it cannot be followed by monitoring macroscopic variables only. It therefore also has to be regarded as nonquasi-static and our result is again in agreement with the second law.

JOULE-THOMSON EXPANSION

In a free expansion no work is done. In a Joule-Thomson expansion the gas is forced into a porous plug at pressure P_2 and emerges at the lower pressure P_1 so that work is done both by and on the gas. If the whole process is adiabatic, we have by (9.13) of Chapter 1,

$$U_1 + P_1 V_1 + \frac{u_1{}^2}{2} = U_2 + P_2 V_2 + \frac{u_2{}^2}{2};$$

and if the difference between initial and final kinetic energies is negligible, we have

$$U_1 + P_1 V_1 = U_2 + P_2 V_2.$$

The function $U + PV$ is called the enthalpy H. Hence

$$H_1 = H_2. \tag{8.7}$$

For an ideal gas, $H = c_P T$. Hence

$$T_1 = T_2$$

and

$$\Delta S = R \ln \frac{V_2}{V_1}. \tag{8.8}$$

Here again the diffusion through the plug cannot be followed macroscopically, the process is not quasi-static, and $\Delta S > 0$.

SHOCK

Now $\Delta\left(U + PV + \dfrac{u^2}{2}\right) = 0$, and the Rankine-Hugoniot relation holds:

$$U_2 - U_1 = \frac{1}{2}(P_1 + P_2)(V_1 - V_2). \tag{8.9}$$

The general expression for the change in entropy is

$$\Delta S = C_V \ln \frac{T_2}{T_1} + R \ln \frac{V_2}{V_1}$$

$$= C_V \ln xy^\gamma \tag{8.10}$$

in the notation of Section 1.10. With the aid of equation (10.15) of that section, ΔS may be expressed in terms of x or y. The entropy change is again positive and for weak shocks is third order in $y - 1 = (V_2 - V_1)/V_1$.

MIXING

Consider a container divided into two equal parts by a wall which separates two different gases, A and B. Suppose that the wall is removed so that A and B both expand to fill the container. Then the increase in entropy associated with the expansion is $2R \ln 2$ (one mol of both A and B). The expansion in this case leads to a mixing and the entropy change associated with the mixing is $2R \ln 2$.

It is clear that there is a mixing no matter how little A differs from B, but that there is no mixing if A is the same as B. If the gases A and B are identical, then removal of the partition leads to no observable change, and therefore the corresponding change in entropy ought to be zero. Since the preceding calculation tells us that in this case ΔS is still $2R \ln 2$, the calculation must be incorrect for the case of identical gases. This inconsistency, which was first noticed by Gibbs, will be further discussed in Chapter 8.

The second law was obtained as an experimental induction. It was of course suggested by examples like the preceding [2].

2.9   Reversibility and Quasi-static Processes

We define the environment E of a system A to consist of all matter that is in thermal contact with A and such that the total system $A + E$ is thermally isolated. For the total system we have, according to (7.2),

$$\Delta S \geq 0.$$

A system A may be said to undergo a *reversible* change if the total system

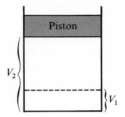

FIGURE 2.1
Gas expands irreversibly from V_1 to V_2.

$A + E$ may be restored to its initial state. Consider a nonquasi-static change of the total system. Then by (7.2)

$$\Delta S_{A+E} > 0.$$

There is now, according to (7.2), no change that will restore $A + E$ to its initial state, for a quasi-static change will leave S unchanged, and any other kind of process will increase S still more. For this reason we may describe processes that are not quasi-static as irreversible.

We shall in addition consider only such causes of irreversibility as stem from the second law. It is therefore possible to regard all quasi-static processes as reversible just as we must consider all other changes as irreversible. Then quasi-static and reversible are equivalent terms. Reversibility therefore means step-by-step as well as global reversibility.

The notion of infinitely slow is sometimes substituted for quasi-static or reversible. However the terms are not equivalent since, for example, an infinitely slow diffusion is not reversible or quasi-static. As was pointed out earlier, a diffusion can be followed only by keeping track of individual molecules and therefore can not be described in terms of macroscopic coordinates or a sequence of thermodynamic states, even though it may occur extremely slowly [3].

We illustrate the concept of reversibility with the following example:

Suppose a gas has undergone an adiabatic free expansion or a diffusion (Figure 2.1) from V_1 to V_2. Suppose we try to restore the gas and its environment to their initial state by letting the piston shown in the figure slowly fall under gravity. This will restore the gas to its original volume but will also heat the gas since the walls are adiabatic and it will also alter the original environment (position of the piston). We may now use the heat so obtained to drive a cyclic engine which will raise the piston. As we shall see, however, the efficiency of this engine must be less than unity; hence this procedure will fail to restore the piston to its original position, unless the gas is cooled below its initial temperature. Therefore the gas and its environment are not both restored to the initial state.

2.10   Kelvin and Clausius Formulations

The preceding example leads naturally to the Kelvin statement of the second law. We shall also give the equivalent Clausius formulation.

A transformation whose only final result is to transform into work heat extracted from a source which is at the same temperature throughout is impossible. (Kelvin)

A transformation whose only final result is to transfer heat from a body at a given temperature to a body at a higher temperature is impossible. (Clausius)

The Kelvin and Clausius formulations and their relationship to other formulations of the second law are discussed in Appendix A [4].

Notice that the Clausius statement refers to the concept "higher temperature." With only the zero and the first law it is not even possible to say what we mean by a higher temperature. Although the zero law defines a temperature order, as well as a temperature equality, it leaves the designation "higher" as a matter of convention, and the first law has nothing to add to this point.

However, when two systems of different temperatures come to thermal equilibrium by thermal conduction for example [4], it is in fact observed that the energy flow is always in the same direction; i.e., if A is in equilibrium with B, and A' is in equilibrium with B', then if heat flows from A to A', it does also from B to B'. We shall say that the system that is losing energy is at the higher temperature.

2.11   The Second Law and Temperature Order

The fact that there is a spontaneous direction of heat flow provides both a definition of higher temperature and a formulation of the second law as follows.

If the systems $A(t_1)$ and $B(t_2)$ come to thermal equilibrium without producing any other change (so that the total system is adiabatically enclosed and does no work), then the sign of ΔQ_A is independent of A and B and of the equalization process and depends only on (t_1, t_2) [5]. If $\Delta Q_A < 0$, we say $t_1 > t_2$.

For if $\Delta Q_A < 0$ in one process, then if $\Delta Q_A > 0$ according to any other process, the second process would provide a perfect refrigerator and so violate the Clausius statement as previously formulated.

It is possible to base the development of thermodynamics on the second law in this form. If one wants to emphasize these ideas, one may say that the zero law establishes the existence of temperature equality and order; the first law defines heat, and the second law relates temperature order to a heat measurement; at the same time the second law establishes the compatibility relation between the thermal and caloric equations that follow from the zero and first law respectively.

2.12 Carathéodory Statement of Second Law

According to Clausius and Kelvin, certain thermal processes are forbidden. Carathéodory has given a similar formulation of the second law [6], but one that is logically more economical, namely:

Arbitrarily near to any given state are other states which cannot be reached from the initial state by an adiabatic process.

The adiabatic process referred to here may or may not be quasi-static. Consider the quasi-static case first.

QUASI-STATIC PROCESS

We begin with the two dimensional case. According to the Carathéodory statement, it is possible to find a point R which cannot be reached from point P by an adiabatic process (Figure 2.2).

In the two dimensional case we have

$$dQ = X_1\,dx_1 + X_2\,dx_2 = 0 \qquad (12.1a)$$

or

$$\frac{dx_2}{dx_1} = \frac{-X_1}{X_2}. \qquad (12.1b)$$

There is a single integral curve passing through P. If R does not lie on this curve, then R is adiabatically inaccessible from P. Since such an R may always be found, the Carathéodory statement is not a physical statement about the two dimensional case. Similarly we found that the first formulation (existence of integrating factor) told nothing about the case $n = 2$, since an integrating denominator always exists in that case, also. The second law is of course empty for $n = 2$ in whatever form it is stated.

In the general case $(n > 2)$ we shall now prove the theorem

$$(a) \underset{\longrightarrow}{\overset{\longleftarrow}{}} (b) \qquad (12.2)$$

FIGURE 2.2
The point R is not accessible from point P.

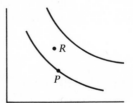

38

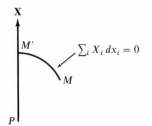

$$\sum_i X_i \, dx_i = 0$$

FIGURE 2.3
Plane MPX determined by M and vector X at P.

where (a) means: there are inaccessible points;
(b) means: there is an integrating denominator.

Before proving this result let us make two simple geometrical remarks, namely:

$$dx^\mu = X^\mu \, ds \qquad (12.3)$$

defines a curve, parallel to X^μ, and

$$\sum_\mu X^\mu \, dx_\mu = 0 \qquad (12.4)$$

defines a surface element normal to X^μ.

Proof that (a) $\longrightarrow$ (b) for case $n = 3$

Consider the point P (Figure 2.3). In any infinitesimal neighborhood of P we may find an inaccessible point M according to assumption (a).

Let X be the vector field defined by $dQ \, (= \sum_i X_i \, dx_i)$ at P. Pass a plane through M and X. On this plane define a curve passing through M by the equation

$$dQ = \sum_i X_i \, dx_i = 0. \qquad (12.4a)$$

Let the curve intersect the extension of X at M'. It follows that M' is also inaccessible from P. Otherwise M could be reached from P via M'.

Next draw a closed plane curve $\mathcal{C}$ through P and consider the surface composed of curves $(\mathcal{L})$ defined by (12.3) and the vector field X, and intersecting $\mathcal{C}$ (Figure 2.4). This surface resembles a deformed cylinder with generators $\mathcal{L}$ moving around the closed curve $\mathcal{C}$.

Finally, on this deformed cylinder draw the solution curve (Γ)

$$dQ = \sum_i X_i \, dx_i = 0 \qquad (12.4b)$$

through P. Then the curve Γ crosses every curve $(\mathcal{L})$ orthogonally and returns to cut the original curve through P at the point P'.

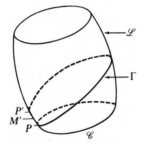

FIGURE 2.4
The solution curve Γ must be a closed curve.

Suppose now that P and P' do not coincide. Then by expanding or contracting the curve $\mathcal{C}$, the point P' may be made to sweep over M'. But if that is possible, then M' is accessible from P and our original hypothesis about the inaccessibility of M' would be contradicted. Hence one must conclude that P and P' do coincide, or that Γ is a closed curve.

Then by Stokes' theorem

$$\oint_\Gamma \mathbf{X}\, d\mathbf{x} = \int_\Gamma \mathrm{curl}_n\, \mathbf{X}\, dS.$$

But for any solution curve, Γ, according to (12.4b),

$$\oint \mathbf{X}\, d\mathbf{x} = 0.$$

Therefore if Γ is an infinitesimal closed curve,

$$(\Delta S)\, \overline{(\hat{\mathbf{n}}\, \mathrm{curl}\, \mathbf{X})} = 0$$

where $\hat{\mathbf{n}}$ is normal to the surface element (ΔS) and hence parallel to $\mathbf{X}$. It follows that

$$\mathbf{X}\, \mathrm{curl}\, \mathbf{X} = 0. \tag{12.5}$$

This last is the condition (2.7) for the existence of an integrating denominator. In the case of a general Pfaffian ($n > 3$), the same argument goes through if we understand by the above equations the n-dimensional generalization of Stokes' theorem.

Proof that $(b) \longrightarrow (a)$

This result follows from the fact that the existence of an integrating factor implies the existence also of a family of nonintersecting integral surfaces. The existence of inaccessible points is then clear.

NONQUASI-STATIC CASE

Let us return to a two dimensional system (Figure 2.5). If the adiabatic process is not quasi-static, it is possible to move off the isentropic lines.

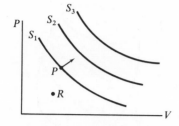

The point R is inaccessible from point P even if adiabatic process is not quasi-static.

However, according to our previous formulations of the second law we may move off only in the direction of increasing entropy (arrow). Hence there are inaccessible points on the low entropy side.

We now turn the argument around and deduce the principle of the increase of entropy from the Carathéodory principle.

Consider an adiabatically enclosed system of two bodies (A and B) in thermal contact. Equilibrium can be characterized by (V_a, V_b, S). Consider the changes

$$(V_a, V_b, S) \xrightarrow{\alpha} (V_a', V_b', S) \xrightarrow{\beta} (V_a', V_b', S').$$

The change (α) is supposed to be adiabatic and quasi-static; (β) is supposed to be adiabatic also, but not quasi-static. If $S' - S$ were positive in some processes and negative in others, it would be possible by combining positive and negative changes to reach final states arbitrarily close to the initial state and thus violate the Carathéodory postulate. Hence we conclude from this postulate that $\Delta S > 0$ always or $\Delta S < 0$ always. By appealing to any single experiment we decide between the two possibilities.

There is one arbitrary element here: the absolute temperature could be chosen to have a negative sign—it would only be necessary to assign a negative temperature to the triple point of water. Since we have

$$\Delta S = \frac{\Delta Q}{T},$$

S would also change sign. The usual convention of course is to take T positive and $\Delta S > 0$.

We mentioned earlier the possibility of putting the zeroth and second law into close correspondence by emphasizing the concept of temperature order. They may also be put into correspondence by formulating the zeroth law in terms of inaccessibility as follows:

Arbitrarily close to any given state are other states that cannot be reached by a process that maintains thermal contact (equilibrium) with the given state.

If we substitute "thermal isolation" for "thermal contact" the preceding statement of the zeroth law goes into the Carathéodory statement of the second law. These two statements guarantee that the thermodynamic states may be arranged in families of nonintersecting hypersurfaces, either isothermal or isentropic.

Notes and References

1. The following is a geometrical counter example. The vector **X** defines a curve:

$$dx_1 = X_1\, dS$$
$$dx_2 = X_2\, dS$$
$$dx_3 = X_3\, dS.$$

Suppose **X** represents a helix. Then

$$
\begin{array}{lll}
X_1 = -x_2 & dx_1 = -x_2\, dS & \text{curl}_1 X = 0 \\
X_2 = x_1 & dx_2 = x_1\, dS & \text{curl}_2 X = 0 \\
X_3 = k & dx_3 = k\, dS & \text{curl}_3 X = 2.
\end{array}
$$

In this case **X** curl **X** $= 2k \neq 0$ and the corresponding Pfaffian has no integrating factor unless the pitch of the helix vanishes.

2. In all of these examples the intermediate states can be described only microscopically. There is in each case, moreover, a simple kinetic description, namely, an expansion in phase space: in (a) and (b) in only the configurational and momentum parts of phase space, respectively. In general (c) and (d) involve simultaneous expansions in both parts of phase space, although (c) for the perfect gas is not different from (a). In (e) the "expansion" is caused by the mixing.

3. It is also perhaps worth mentioning that thermodynamic reversibility refers only to a sequence of macroscopic states which are labeled by hydrodynamic and thermodynamic variables. (For example we exclude the motion of an undamped pendulum, which is reversible but does not consist of a sequence of equilibrium states in the preceding sense, since one is here dealing with the degrees of freedom of a rigid body rather than with hydrodynamic and thermodynamic variables.)

4. The analysis of the Kelvin and Clausius formulations described in Appendix A follows that of E. Fermi, *Thermodynamics* (Dover, New York, 1956).

5. If A and B are constrained from doing work, then $\Delta Q = \Delta U$. We may assume complete knowledge of U on the basis of formula (6.2) of Chapter 1. Then ΔU and hence ΔQ is determined.

6. The Carathéodory formulation is discussed in references [1], [2], and [3] of Chapter 1. See also A. B. Pippard, *The Elements of Classical Thermodynamics* (Cambridge Univ. Press, Cambridge, 1957).

One-component Systems Phase Transitions and Low Temperatures

3.1 Qualitative Description of a One-component System

The discussion of the first and second laws has so far been very general. Let us next specialize these general considerations to the simplest realistic systems, namely those with only two degrees of freedom; for example, chemically pure systems which do only pressure work. These are known as one-component systems and differ among themselves in their thermal and caloric equations of state and in the nature of their phase transitions. If the system is, for example, a mass of hydrogen, it may exist in three kinds of phase: solid, liquid, and gas; but if it is helium, then there is an additional superfluid phase. In the most general case the complete thermodynamic space will be divided into domains separated by surfaces across which there are discontinuities or other nonanalytic behavior in some of the thermodynamic functions.

The following remarks may be made about the general phase diagram that is illustrated in Figure 3.1.

(a) As V and/or T become very large, the system behaves like a perfect gas. Classical mechanics holds very well in this limit.

(b) According to the second law, the $T = 0$ isotherm is singular. The behavior of matter in the neighborhood of absolute zero is governed by the third law (which we consider in the next chapter). Quantum mechanics becomes very important here.

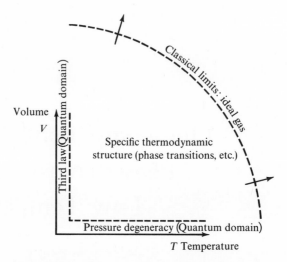

FIGURE 3.1
Generic phase diagram of a one-component system.

(c) As V is decreased to the high density limit, one again moves into a quantum regime. The white dwarf stars are examples of very dense matter and illustrate quantum degeneracy.

(d) The second law establishes a compatibility relation between the thermal and caloric equations that hold in the separate domains of the distinct phases and it also imposes conditions on the phase boundaries.

According to (a), (b), and (c), all forms of matter satisfy certain asymptotic conditions. In these limits one sees behavior that is relatively simple and universal.

In the central region there is still a great diversity in thermodynamic structure, diversity that is allowed by the conditions (d). We shall now consider these conditions.

3.2 Thermodynamic Relations Holding for Simple Systems

In this section we shall record some of the formal relations that follow from the second law and hold for simple systems. They are generalizations of relations given in Chapter 1 for matter approximating the behavior of a perfect gas.

The first and second laws may be combined in the form

$$T \, dS = dU + P \, dV. \tag{2.1}$$

The integrability condition for dS is

$$T \left(\frac{\partial P}{\partial T} \right)_V = \left(\frac{\partial U}{\partial V} \right)_T + P. \tag{2.2}$$

Although it is possible to write the preceding equation for any system with only two degrees of freedom, it should be remembered that the universal significance of the T appearing therein can be established only by considering systems with more degrees of freedom. These two equations, (2.1) and (2.2), may again be combined:

$$T \, dS = \left[\left(\frac{\partial U}{\partial V} \right)_T + P \right] dV + \left(\frac{\partial U}{\partial T} \right)_V dT$$

$$dS = \left(\frac{\partial P}{\partial T} \right)_V dV + \frac{1}{T} \left(\frac{\partial U}{\partial T} \right)_V dT. \tag{2.3}$$

The specific heat at constant volume, C_V, is

$$C_V = T \left(\frac{\partial S}{\partial T} \right)_V = \left(\frac{\partial U}{\partial T} \right)_V. \tag{2.4}$$

Hence

$$dS = C_V \frac{dT}{T} + \left(\frac{\partial P}{\partial T} \right)_V dV. \tag{2.5}$$

The integrability condition for dS in this form reads

$$\frac{1}{T} \left(\frac{\partial C_V}{\partial V} \right)_T = \left(\frac{\partial^2 P}{\partial T^2} \right)_V. \tag{2.6}$$

There is a corresponding set of relations involving the *enthalpy:*

$$H = U + PV$$

beginning with

$$T \, dS = dH - V \, dP. \tag{2.7}$$

The other relations may be obtained by the substitutions

$$U \longrightarrow H, \qquad P \longrightarrow -V, \qquad \text{and} \qquad V \longrightarrow P.$$

Hence the analogues of the relations (2.2)–(2.6) are

$$-T \left(\frac{\partial V}{\partial T} \right)_P = \left(\frac{\partial H}{\partial P} \right)_T - V \tag{2.8}$$

$$C_P = \left(\frac{\partial H}{\partial T} \right)_P \tag{2.9}$$

$$dS = - \left(\frac{\partial V}{\partial T} \right)_P dP + C_P \frac{dT}{T} \tag{2.10}$$

$$\frac{1}{T}\left(\frac{\partial C_P}{\partial P}\right)_T = -\left(\frac{\partial^2 V}{\partial T^2}\right)_P.$$ (2.11)

For an arbitrary change with specific heat, C_a,

$$C_a\, dT = T\left(\frac{\partial P}{\partial T}\right)_V dV + C_V\, dT$$ (2.12)

$$C_a - C_V = T\left(\frac{\partial P}{\partial T}\right)_V \left(\frac{\partial V}{\partial T}\right)_a.$$ (2.13)

Similarly,

$$C_P - C_a = T\left(\frac{\partial V}{\partial T}\right)_P \left(\frac{\partial P}{\partial T}\right)_a.$$ (2.14)

In particular,

$$C_P - C_V = T\left(\frac{\partial P}{\partial T}\right)_V \left(\frac{\partial V}{\partial T}\right)_P$$ (2.15)

$$C_P - C_V = T\, V\, \beta^2 / K_T$$ (2.16)

where

$$\beta = \frac{1}{V}\left(\frac{\partial V}{\partial T}\right)_P$$ (2.17)

$$K_T = -\frac{1}{V}\left(\frac{\partial V}{\partial P}\right)_T.$$ (2.18)

Since $K_T > 0$, it follows that

$$C_P - C_V > 0.$$ (2.19)

Also,

$$\frac{C_P - C_a}{C_V - C_a} = \frac{-(\partial V/\partial T)_P\,(\partial P/\partial T)_a}{(\partial P/\partial T)_V\,(\partial V/\partial T)_a}$$

$$= \frac{K_T}{K_a}.$$ (2.20)

If $C_a = 0$, we have

$$\frac{C_P}{C_V} = \frac{K_T}{K_S}.$$ (2.21)

Since $C_P - C_V > 0$, it follows that $C_P/C_V > 1$ and therefore by (2.21) that the isentropics are steeper than the isothermals. This conclusion depends on the experimental fact $K_T > 0$. The theoretical basis of this and similar remarks depends on stability considerations which will be discussed in Chapter 4.

3.3   Minimal Thermodynamic Description

The existence of the entropy implies that the caloric and thermal equations of state are not independent. For example,

$$T\left(\frac{\partial P}{\partial T}\right)_V - P = \left(\frac{\partial U}{\partial V}\right)_T \tag{3.1}$$

$$T\left(\frac{\partial^2 P}{\partial T^2}\right)_V = \left(\frac{\partial C_V}{\partial V}\right)_T. \tag{3.2}$$

Consider now a gas obeying van der Waals' equation:

$$P = -\frac{a}{V^2} + \frac{RT}{V - b}$$

$$\left(\frac{\partial P}{\partial T}\right)_V = \frac{R}{V - b}. \tag{3.3}$$

Hence

$$\left(\frac{\partial U}{\partial V}\right)_T = -P + \frac{RT}{V - b} = \frac{a}{V^2}$$

$$U = -\frac{a}{V} + \phi(T). \tag{3.4}$$

The caloric equation is limited in this way by the second law. The two parts of this expression for U of course correspond to potential and kinetic energies.

Since the thermal and caloric equations are not independent, to assign both may be inconsistent and at best is redundant. One then asks what is the minimal thermodynamic description of a simple substance.

Let us assume that the thermal equation of state is given. How much additional information is necessary to fix U? It is enough to know $(\partial U/\partial V)_T$ and $(\partial U/\partial T)_V = C_V$. The first of these differential coefficients is given by (3.1) while the second may be obtained by integrating (3.2) provided that proper boundary conditions are given, namely, a specification of C_V along any nonisothermal curve.

Thus, the thermodynamic properties of a simple substance may be completely specified by giving its thermal equation of state and the specific heat along any curve that is not an isothermal.

3.4   Other Characteristic Functions

A given thermodynamic state corresponds to definite values of the variables V, P, S, T, U. In addition, the following functions are often of interest:

$$H = U + PV \tag{4.1}$$
$$A = U - TS \tag{4.2}$$
$$G = U + PV - TS. \tag{4.3}$$

The first and second laws may be combined in the following form

$$T \, dS \geq dU + P \, dV \tag{4.4}$$

where the equality sign holds for reversible changes [equation (7.1) of Chapter 2]. In terms of the characteristic functions just introduced

$$dU \leq T \, dS - P \, dV \tag{4.5}$$
$$dH \leq T \, dS + V \, dP \tag{4.6}$$
$$dG \leq - S \, dT + V \, dP \tag{4.7}$$
$$dA \leq - S \, dT - P \, dV. \tag{4.8}$$

In case the equality sign holds, one can deduce the following equations of state:

$$\left(\frac{\partial U}{\partial S}\right)_V = T \qquad \left(\frac{\partial U}{\partial V}\right)_S = - P \tag{4.9}$$

$$\left(\frac{\partial H}{\partial S}\right)_P = T \qquad \left(\frac{\partial H}{\partial P}\right)_S = V \tag{4.10}$$

$$\left(\frac{\partial G}{\partial T}\right)_P = - S \qquad \left(\frac{\partial G}{\partial P}\right)_T = V \tag{4.11}$$

$$\left(\frac{\partial A}{\partial T}\right)_V = - S \qquad \left(\frac{\partial A}{\partial V}\right)_T = - P. \tag{4.12}$$

(For each characteristic function there are two equations: one thermal and the other caloric.) Between these equations of state there are the following conditions of compatibility (Maxwell relations):

$$\left(\frac{\partial T}{\partial V}\right)_S = - \left(\frac{\partial P}{\partial S}\right)_V \tag{4.13}$$

$$\left(\frac{\partial T}{\partial P}\right)_S = \left(\frac{\partial V}{\partial S}\right)_P \tag{4.14}$$

$$-\left(\frac{\partial S}{\partial P}\right)_T = \left(\frac{\partial V}{\partial T}\right)_P \tag{4.15}$$

$$\left(\frac{\partial S}{\partial V}\right)_T = \left(\frac{\partial P}{\partial T}\right)_V. \tag{4.16}$$

In most physical applications it is convenient to choose T and P as independent variables, and therefore to use the *Gibbs* function G. In chemical applications H is frequently of interest, since H represents the heat of reaction at constant pressure. According to (4.8) the decrease in A, called the *free energy*, measures the work done in a reversible isothermal transformation and is an upper bound in the irreversible case.

In Appendix B we describe a systematic classification of all thermodynamic formulas [1].

3:5   Phase Transition of First Order

In general G and its derivatives vary smoothly over the PT-plane. However along certain lines there may exist discontinuities in the first derivatives (V and S). These are called phase transitions of the first kind. It is possible also that V and S, as well as G, are continuous while higher order derivatives of G are discontinuous; these are designated higher order phase transitions and are smoother than first order transitions.

We shall now obtain the equations for a first order transition. By the second law,

$$dG \leq V\,dP - S\,dT. \tag{5.1}$$

Consider changes for which $dP = dT = 0$. Then

$$dG \leq 0 \tag{5.2}$$

where the equality sign holds for the reversible changes. In irreversible changes under these conditions, G will decrease and move toward the minimum value compatible with the constraints on the physical system. It is of course conceivable that no irreversible mechanisms are available for lowering the Gibbs function to the minimum value consistent with these constraints. However we shall always assume that the required mechanisms *are* available or in other words that any transformation, which is not explicitly forbidden, is allowed. Therefore we exclude cases of metastability such as supercooling.

If we have a two-phase system, for example one with specific Gibbs functions g_1 and g_2, we may write for the Gibbs function G and mass M of the total system

$$G = m_1 g_1 + m_2 g_2 \tag{5.3a}$$

$$M = m_1 + m_2. \tag{5.3b}$$

In this case the only constraint is the conservation of mass. If $g_1 < g_2$, G will be decreased if the second phase converts to the first. Then

$$G_{\text{minimum}} = g_1 M. \tag{5.3}$$

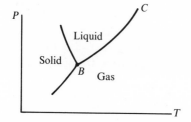

FIGURE 3.2
Phase diagram showing triple point (B) and critical point (C).

Arbitrary amounts of m_1 and m_2 are stable with respect to each other only if $g_1 = g_2$. Then the condition for phase stability is

$$g_1 = g_2. \tag{5.4}$$

If two phases, say ice and water, are put in contact they will in general not be in equilibrium; the phase with the higher g will transform into the other. This change will go on irreversibly until the unstable phase has disappeared.

The differential form of equation (5.4) is

$$dg_1 = dg_2$$

$$\left(\frac{\partial g}{\partial P}\right)_1 dP + \left(\frac{\partial g}{\partial T}\right)_1 dT = \left(\frac{\partial g}{\partial P}\right)_2 dP + \left(\frac{\partial g}{\partial T}\right)_2 dT \tag{5.5}$$

$$\frac{dP}{dT} = \frac{\Delta(\partial g/\partial T)}{\Delta(\partial g/\partial P)}$$

where

$$\Delta(\ ) = (\ )_1 - (\ )_2.$$

Hence

$$\frac{dP}{dT} = \frac{\Delta S}{\Delta V} = \frac{l}{T\Delta V} \tag{5.6}$$

where l is the latent heat of the transition. This is the Clausius-Clapeyron relation. Figure 3.2 illustrates a phase diagram in the neighborhood of the triple point.

The liquid-vapor line ends at a point, the so-called critical point, C, where the two phases have become indistinguishable. Beyond this point there is only a single phase. At C we have $\Delta S = \Delta V = 0$, so that $dP/dT = 0/0$.

In such a situation where first derivatives of G are continuous it is still mathematically possible for second derivatives to be discontinuous. That is not the case at a critical point but is realized in other situations and is described as a phase change of higher order [2-5].

3.6 Phase Transition of Second Order

The phase equilibrium just described may be characterized as follows:

$$\Delta g = 0$$

$$\Delta V \neq 0 \quad \text{or} \quad \Delta \left(\frac{\partial g}{\partial P} \right)_T \neq 0$$

$$\Delta S \neq 0 \quad \quad \Delta \left(\frac{\partial g}{\partial T} \right)_P \neq 0,$$

(6.1)

i.e., g is continuous but its first derivatives are discontinuous. We now consider a higher order transition:

$$\Delta g = 0 \tag{6.2a}$$

$$\Delta \left(\frac{\partial g}{\partial P} \right)_T = 0 \tag{6.2b}$$

$$\Delta \left(\frac{\partial g}{\partial T} \right)_P = 0, \tag{6.2c}$$

but the second derivatives are discontinuous. The continuity of the first derivatives may be expressed as follows:

$$V_1 = V_2 \tag{6.3}$$

$$S_1 = S_2. \tag{6.4}$$

The differential form of equation (6.3) is

$$\Delta \left(\frac{\partial V}{\partial P} \right) dP + \Delta \left(\frac{\partial V}{\partial T} \right) dT = 0 \tag{6.3a}$$

or

$$\frac{dP}{dt} = - \frac{\Delta(\partial V/\partial T)}{\Delta(\partial V/\partial P)} = \frac{\Delta \beta}{\Delta K} \tag{6.3b}$$

where

$$\beta = \frac{1}{V} \left(\frac{\partial V}{\partial T} \right)_P$$

$$K = - \frac{1}{V} \left(\frac{\partial V}{\partial P} \right)_T.$$

For (6.4) we have similarly

$$\Delta \left(\frac{\partial S}{\partial P} \right) dP + \Delta \left(\frac{\partial S}{\partial T} \right) dT = 0$$

$$\frac{dP}{dT} = - \frac{\Delta(\partial S/\partial T)_P}{\Delta(\partial S/\partial P)_T}. \tag{6.4a}$$

We have

$$\left(\frac{\partial S}{\partial T}\right)_P = \frac{C_P}{T}$$

$$\left(\frac{\partial S}{\partial P}\right)_T = -\left(\frac{\partial V}{\partial T}\right)_P.$$

Hence

$$\frac{dP}{dt} = \frac{1}{VT}\frac{\Delta C_P}{\Delta\beta}. \tag{6.4b}$$

Equations (6.3b) and (6.4b) describe phase changes of the second order, and are known as the Ehrenfest equations. In order to apply these equations it is necessary that the discontinuities in C_P and β be finite, but this situation is usually not realized.

For example, a so-called λ-transition (to be described in the next chapter) is similar to the case just considered but ΔC_p is now infinite at the λ-line. Therefore we consider two states (1) and (2) which are close but on opposite sides of the λ-line (see Figure 3.3). Ehrenfest's equations may then be applied in the following form to the immediate neighborhood of the λ-line:

$$C_P(2) - VT\left(\frac{dP}{dT}\right)\beta(2) = C_P(1) - VT\left(\frac{dP}{dT}\right)\beta(1)$$

$$\beta(2) - \left(\frac{dP}{dT}\right)K(2) = \beta(1) - \left(\frac{dP}{dT}\right)K(1).$$

One may hold state (1) [or (2)] fixed, and let (2) [or (1)] vary, and thereby obtain the Pippard equations [6],

$$C_P = VT\left(\frac{dP}{dT}\right)\beta + \text{constant} \tag{6.5}$$

$$\beta = \left(\frac{dP}{dT}\right)K + \text{constant}. \tag{6.6}$$

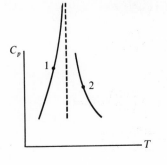

FIGURE 3.3
Specific heat near λ-point.

In the next chapter we shall consider two examples, one illustrating the simple Ehrenfest situation and the other the λ-transition.

3.7 Qualitative Remarks on Phase Transitions

From the microscopic point of view it turns out that the entropy measures the degree of organization of a physical system. In general as the temperature is lowered, matter becomes more highly organized. For example, the molecules of a gas are less correlated than those of a liquid. Liquids in turn are not as highly organized as solids. Usually as the temperature and pressure are varied, the degree of organization changes continuously, but across certain lines in the *PT*-plane one sees discontinuous changes in the state of aggregation.

These discontinuities may be large for a first order transition, or smaller if the transition is of higher order. For example, the solid-liquid transition is first order and abrupt while a solid-solid transition may be higher order and smoother. We mention two examples of the latter.

(1) A solid changes its crystalline structure, i.e., the symmetry group of the crystal changes discontinuously. For example, if the crystal loses some of its symmetry elements, its group degenerates into a subgroup at the transition.

(2) A solid changes its magnetic structure (a Curie transition) but its crystalline structure is undisturbed.

The change in organization accomplished by the transition is associated with nonanalytic behavior of the thermodynamic functions. The theoretical problem is to measure the difference in symmetry of the separate phases by means of these functions. Although the solution to this problem is understood in principle and in simple systems, the application needed here is very complex [7].

In order to deal with higher order transitions systematically, Ehrenfest proposed a classification according to which the order of the transition is determined by the lowest order derivative of the Gibbs function that shows a discontinuity. Unfortunately this simple classification has not turned out to be very relevant since the observed discontinuities are usually infinite. For more realistic discussions see references [2-6] and [8] at the end of this chapter.

3.8 Empirical Determination of the Absolute Temperature Scale

In Section 2.6 an approximate realization of the absolute temperature scale was described. We shall now discuss an exact realization.

The Joule-Kelvin coefficient of an ideal gas vanishes as we have seen. For a real gas, however,

$$\left(\frac{\partial t}{\partial P}\right)_H \neq 0 \tag{8.1}$$

and measurement of this coefficient may be used to fix the absolute temperature scale, with arbitrary precision.

According to (6.1) of Chapter 2,

$$\ln \frac{T_2}{T_1} = \int_{t_1}^{t_2} \frac{(\partial P/\partial t)_V \, dt}{(\partial U/\partial V)_t + P}. \tag{8.2}$$

By the substitutions $U \longrightarrow H$, $P \longrightarrow -V$, and $V \longrightarrow -P$ [see equation (2.7)], equation (8.2) goes into

$$\ln \frac{T_2}{T_1} = \int_{t_1}^{t_2} \frac{-(\partial V/\partial t)_P \, dt}{(\partial H/\partial P)_t - V}. \tag{8.3}$$

But

$$\left(\frac{\partial H}{\partial P}\right)_t \left(\frac{\partial P}{\partial t}\right)_H \left(\frac{\partial t}{\partial H}\right)_P = -1.$$

Therefore

$$\ln \frac{T_2}{T_1} = \int_{t_1}^{t_2} \frac{(\partial V/\partial t)_P \, dt}{C_P(\partial t/\partial P)_H + V} = I(t_2, t_1) \tag{8.4}$$

and if $(\partial t/\partial P)_H$ is measured, $I(t_2, t_1)$ may be computed. For real gases the Joule-Kelvin coefficient is relatively small, i.e.,

$$C_P \left(\frac{\partial t}{\partial P}\right)_H \ll V. \tag{8.5}$$

Hence

$$I(t_2, t_1) \cong \int_{t_1}^{t_2} \frac{1}{V} \left(\frac{\partial V}{\partial t}\right)_P dt$$

$$\cong \alpha(t_2 - t_1) \tag{8.6}$$

where α is the average thermal expansion. Then by (8.4)

$$\ln \frac{T_2}{T_1} \cong \frac{T_2 - T_1}{T_1} \cong \alpha(t_2 - t_1).$$

Identifying T_1 and T_2 with the ice and steam points and choosing $T_2 - T_1 = t_2 - t_1 = 100$, one finds

$$T_1 \cong \frac{1}{\alpha} \cong 273°K.$$

For an exact determination it is of course necessary to use (8.4). The fact that (8.5) and (8.6) become better approximations at large V and/or large T is an illustration of the approach to perfect gas behavior under these conditions.

The above method cannot be extended to very low temperatures where all forms of matter condense. In fact $T = 0$ is a singular hypersurface and the physical domain near $T = 0$ therefore demands special attention. We shall now consider the very low temperature domain.

3.9 Experimental Realization of Very Low Temperatures and the Joule-Kelvin Effect

Temperatures of liquid helium are usually reached by performing in sequence:
(a) a reversible adiabatic expansion,
(b) a Joule-Kelvin expansion. [9]
Finally by the use of adiabatic demagnetization, temperatures of the order of 10^{-6} K may be attained. [10]

According to equation (2.10), the first step is described by

$$0 = c_P \, dT - T \left(\frac{\partial V}{\partial T} \right)_P dP. \tag{9.1}$$

Since $(\partial V/\partial T)_P > 0$, step (a) always leads to a cooling. However, as the temperature is lowered, this method becomes less effective.

We now consider the Joule-Kelvin effect. The general situation is illustrated in the Figure 3.4 showing curves of constant enthalpy. The maxima $(\partial T/\partial P)_H = 0$ define the inversion curve. The initial and final states of the gas must lie on the same H-curve but the process connecting them can not be shown since it is not quasi-static. If both points lie to the left of the inversion curve, the gas will cool in a Joule-Kelvin expansion; if they lie to the right, the gas will warm.

By assuming van der Waals' equation for the gas, one may arrive at some rough conclusions. We have in general

$$\left(\frac{\partial T}{\partial P} \right)_H = \frac{1}{c_P} \left[T \left(\frac{\partial V}{\partial T} \right)_P - V \right] \tag{9.2}$$

$$= -\frac{1}{c_P} \frac{1}{(\partial P/\partial V)_T} \left[T \left(\frac{\partial P}{\partial T} \right)_V + V \left(\frac{\partial P}{\partial V} \right)_T \right]. \tag{9.3}$$

FIGURE 3.4
Curves of constant enthalpy
and inversion curve.

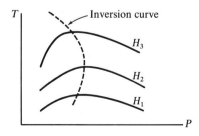

From van der Waals' equation, (3.3),

$$\left(\frac{\partial P}{\partial V}\right)_T = \frac{2a}{V^3} - \frac{RT}{(V-b)^2}$$

$$\left(\frac{\partial P}{\partial T}\right)_V = \frac{R}{V-b}$$

$$\left(\frac{\partial T}{\partial P}\right)_H = -\frac{1}{C_P}\frac{1}{(\partial P/\partial V)_T}\left[\frac{2a}{V^2} - \frac{RTb}{(V-b)^2}\right]$$

$$\cong -\frac{1}{C_P}\frac{1}{(\partial P/\partial V)_T}\frac{Rb}{V^2}\left\{\frac{2a}{Rb} - T\right\}$$

(9.4)

which holds for $V \gg b$. The sign of the Joule-Kelvin coefficient is determined by the sign of the curly bracket, since the coefficient of the bracket is positive. From this last equation one gets an estimate of the maximum inversion temperature, namely

$$T^* = \frac{2a}{Rb}.$$

(9.5)

If T^* is greater than the initial temperature of the gas, it is possible to choose the initial pressure so that cooling will take place. If T^* is less than the initial temperature, the gas will be heated by a Joule-Kelvin expansion. Hence to use this method for cooling, the gas first must be brought below the maximum inversion temperature T^*.

T^* is evidently high for chemically active gases (large a) and low for inert gases; for example,

$$T^*(\mathrm{H}) > T^*(\mathrm{He}).$$

Cooling to low temperatures is accomplished in the helium liquefier by first letting the helium do adiabatic work until it cools below its inversion temperature, and then letting it undergo a Joule-Kelvin expansion. By repeated expansions, preceded by counterflow through a heat exchanger, liquid helium may be produced.

3.10   Adiabatic Demagnetization

By pumping away the helium vapor, one may get to temperatures of $1°\mathrm{K}$ with He^4 and $.35°\mathrm{K}$ with He^3. To go lower, one may employ a procedure depending on the adiabatic demagnetization of a paramagnetic salt.

A paramagnetic substance may be described by the equation

$$B = H + 4\pi M$$

(10.1)

where M vanishes if H does. We also write, for weak fields,

$$B = \mu H \tag{10.1a}$$

$$M = \chi H \tag{10.1b}$$

where μ and χ are called the magnetic permeability and susceptibility. Hence also

$$\mu = 1 + 4\pi\chi. \tag{10.1c}$$

When the magnetization of a paramagnetic salt is increased by dM the work done by the salt is $-H\,dM$. The first law is then

$$T\,dS = dU - H\,dM. \tag{10.2}$$

The U appearing here contains the field energy [**11**].
All the previous thermodynamic formulae are applicable here after the substitutions:

$$P \longrightarrow -H \tag{10.3a}$$

$$V \longrightarrow M. \tag{10.3b}$$

For example, by (2.5)

$$T\,dS = C_M\,dT - T\left(\frac{\partial H}{\partial T}\right)_M dM \tag{10.4}$$

and by (2.10) and (2.15)

$$T\,dS = C_H\,dT + T\left(\frac{\partial M}{\partial T}\right)_H dH \tag{10.5}$$

$$C_H - C_M = -T\left(\frac{\partial H}{\partial T}\right)_M \left(\frac{\partial M}{\partial T}\right)_H, \tag{10.6}$$

etc.

ISOTHERMAL MAGNETIZATION

In an isothermal process we have by (10.5)

$$dQ = T\left(\frac{\partial M}{\partial T}\right)_H dH. \tag{10.7}$$

Empirically it is found that

$$\left(\frac{\partial M}{\partial T}\right)_H < 0 \tag{10.8}$$

and hence

$$dQ < 0. \tag{10.9}$$

Therefore if the magnetic field is increased, the specimen will remain at constant temperature only if it gives out heat.

ADIABATIC DEMAGNETIZATION

If the magnetic field is changed slowly and adiabatically we have by (10.5)

$$0 = C_H \, dT + T \left(\frac{\partial M}{\partial T} \right)_H dH, \tag{10.10}$$

then

$$\frac{dT}{T} = -\frac{1}{C_H} \left(\frac{\partial M}{\partial T} \right)_H dH. \tag{10.11}$$

As we have just remarked $(\partial M/\partial T)_H < 0$. Hence if $dH < 0$, then $dT < 0$ also.

Cooling can be accomplished (Figure 3.5) by the following two step procedure:

(a) isothermal magnetization,
(b) isentropic demagnetization.

The sample is first magnetized in contact with a heat bath to which it loses energy during the magnetization. Thermal contact is then broken and the magnetic field switched off. If one starts with a paramagnetic salt at a temperature of about 1°K, one can in this way reach, under favorable circumstances, temperatures of the order of 10^{-3} K.

The possibility of carrying out this procedure depends on the sign of $(\partial T/\partial M)_S$. This sign can be related, according to the argument just given, to the sign of $(\partial M/\partial T)_H$, which we take from experiment, or atomic theory.

It is possible however to determine the correct sign by appealing only to the interpretation of entropy as a measure of disorder and without introducing any specific atomic models. The argument depends on the following:

$$\left(\frac{\partial S}{\partial T} \right)_H > 0 \tag{10.12}$$

$$\left(\frac{\partial S}{\partial H} \right)_T < 0, \tag{10.13}$$

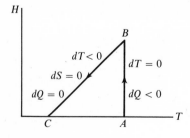

FIGURE 3.5
Adiabatic demagnetization. *AB* is isothermal magnetization. *BC* is isentropic demagnetization.

i.e., if S is a measure of disorder in the spin system, it is decreased by aligning spins—which may be accomplished either by increasing the magnetic field or by lowering the temperature. However,

$$\left(\frac{\partial S}{\partial T}\right)_H \left(\frac{\partial T}{\partial H}\right)_S \left(\frac{\partial H}{\partial S}\right)_T = -1. \tag{10.14}$$

Hence

$$\left(\frac{\partial T}{\partial H}\right)_S > 0. \tag{10.15}$$

CALIBRATION OF TEMPERATURE SCALE

The problem of extending the temperature scale into the regions accessible by this technique may be handled in the following way.

Assume that the absolute temperature is experimentally defined at and above the temperature T_i. Then it is possible to determine S at this temperature as a function of H by the equation

$$dS = \left(\frac{\partial M}{\partial T}\right)_H dH. \tag{10.16}$$

Let the lines $11'$, $22'$, $33'$, ... in Figure 3.6 be isentropics. Then

$$S(i') = S(i).$$

But the points i' represent final states for adiabatic demagnetizations from 1, 2, 3, ... Let some empirical temperature t be measured at these final states. Let $S(i')$ be plotted against $t(i')$ in Figure 3.7. We have

$$dQ = T\,dS$$

$$\frac{dQ}{dt} = T\frac{dS}{dt}. \tag{10.17}$$

From the St plot of Figure 3.7, it is possible to determine the slope dS/dt. It is also possible by an independent experiment to determine the specific heat

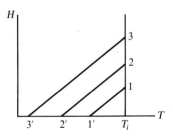

FIGURE 3.6
Entropy at low temperature related to entropy at high field.

$S(i')$

$t(i')$

FIGURE 3.7
Entropy as function of final
empirical temperature.

$c* = dQ/dt$ referred to the empirical temperature t. From (10.17) and these measurements the Kelvin temperature may be determined:

$$T = \frac{c*}{(dS/dt)}.$$ (10.18)

Although the choice of t is not important as a matter of principle, it is convenient to use the so-called Curie temperature, defined by A/χ where χ is the magnetic susceptibility and A is a constant.

MAGNETIC REFRIGERATOR

The adiabatic demagnetization just described leads to a final state of very low temperature. However the final temperature then begins to increase.

It is possible to maintain the low final temperature by a magnetic refrigerator operating between T_i and T_f. The Carnot refrigeration cycle is shown in Figure 3.8. The paramagnetic salt is magnetized in contact with T_i and gives up heat. It is demagnetized in contact with T_f and absorbs heat. It thus repeatedly transfers energy from the low to the high temperature [12].

NUCLEAR SPINS

The temperatures attainable by adiabatic demagnetization are finally limited by cooperative effects among the electronic spins. However, it is possible to cool a system of *nuclear* spins to temperatures of the order of 10^{-6} K by using the paramagnetic salt (the electronic spin system) as a heat sink, and performing the adiabatic demagnetization on the nuclear spin system itself. Again the

FIGURE 3.8
Magnetic refrigerator.

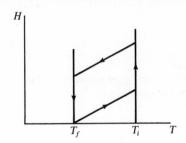

H

T_f　　T_i　　T

cooling is ultimately limited by cooperative effects and the third law. The role of the third law will be discussed in Chapters 4 and 8.

Notes and References

1. P. W. Bridgman, *A Condensed Collection of Thermodynamic Formulas* (Harvard University Press, Cambridge, 1926).
2. For classification and discussion of higher order phase transitions, see A. B. Pippard, *The Elements of Classical Thermodynamics* (Cambridge Univ. Press, Cambridge, 1957), and the following references [3], [4], and [5].
3. H. B. Callen, *Thermodynamics* (Wiley, New York, 1960).
4. L. D. Landau and E. M. Lifshitz, *Statistical Physics* (Pergamon, London, 1958).
5. L. Tisza, *Generalized Thermodynamics* (M.I.T. Press, Cambridge, 1966).
6. Pippard, reference [2] p. 143.
7. In principle this problem should be studied first by relating the quantum mechanical energy spectrum (including in particular its degeneracies) to the symmetry of the corresponding phase, and second by relating the thermodynamic functions to this spectrum by the general principles of statistical mechanics.
8. Kadanoff et al., *Rev. Mod. Phys.* 39: 395 (1967).
9. M. W. Zemansky, *Heat and Thermodynamics* (McGraw-Hill, New York, 1957).
10. K. Mendelssohn, *Cryophysics* (Interscience, New York, 1960).
11. This relation may also be written in the form

$$T\,dS = dU + M\,dH.$$

Now U does not contain the field energy. See for example C. Kittel, *Elementary Statistical Physics* (Wiley, New York, 1958, p. 77).
12. The technical problem of finding a thermal switch may be solved by using lead rods; these are poor heat conductors in the superconducting state and good heat conductors in the normal state. Since application of a suitable magnetic field causes a transition out of the superconducting state, switching the field on and off makes and breaks thermal contact with the reservoir. See reference [9], p. 359.

The Third Law and the Behavior of Matter Near Absolute Zero

4.1 Qualitative Behavior of Matter Near Absolute Zero

The second law defines absolute temperature and entropy and at the same time, by introducing a logarithmic temperature scale, establishes $T = 0$ as a singular surface. The third law completes the formal foundations of thermodynamics by providing a boundary condition on this singular surface in terms of an empirical induction describing the behavior of matter at low temperatures. The general fact recorded in the third law is that the isentropic and isothermal surfaces approach coincidence as the temperature tends toward zero, and the postulate known as the third law states that the two families actually coincide at absolute zero. The observed qualitative behavior is that the entropy tends to decrease as the temperature is lowered, provided the other variables are held constant, until, as absolute zero is approached, the value approached by the entropy is independent of all other variables and may be equated to zero.

Since from the statistical viewpoint the entropy of a physical system measures the "disorder" of that system, the trend summarized by the third law may be described as the tendency of all matter to become more highly organized as the temperature is lowered. This organization may go on continuously in a single phase, as dictated by the thermal and caloric equations of state, or it may occur discontinuously in a phase change.

The mathematical "disorder" is a measure of the number of quantum mechanical states needed to specify the thermodynamics of the given system. Therefore this increase in organization means that the effective number of contributing quantum mechanical states tends to decrease as the temperature is lowered. Finally in the neighborhood of absolute zero, the behavior of the macroscopic physical system may depend on a relatively small number of quantum states; behavior at this point may exhibit macroscopic coherence features that are entirely mysterious from the classical viewpoint. The most dramatic examples of this sort so far observed are superconductivity and superfluidity, which are associated with the two kinds of statistics, Fermi-Dirac and Einstein-Bose, respectively. Other examples of macroscopic quantum behavior should be expected as the temperature is lowered, on the basis of these general arguments.

In other words the third law leads one to expect that all physical systems will be dominated by quantum mechanics as the temperature approaches absolute zero. Therefore we should not regard the known superfluids as anomalous, but rather as examples of highly ordered states which are precursors of the completely ordered states characteristic of absolute zero.

In order to discuss superfluidity we shall first give a little background material about liquid helium [**1, 2, 3**].

4.2   Phase Diagram of Liquid Helium

At $T \cong 2.2°$ and a pressure of 1 atmosphere the specific heat appears to become infinite [**4**]. This point has been called the λ-point because of the appearance of the specific heat curve (Figure 4.1).

The phase diagram of helium is illustrated in Figure 4.2. The line BA is the λ-line, across which there is a discontinuity in C_P. It divides liquid helium into two phases, I and II. The solid and vapor phases are shown as well. (At atmospheric pressure, helium exists as a liquid down to absolute zero; at that temperature it can be solidified only under a pressure of about 25 atm.) Here A and B are triple points; C is a critical point. The transition AB exists only in He^4, not in He^3, and therefore depends in an essential way on the statistics.

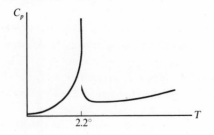

FIGURE 4.1
Specific heat near λ-point of helium.

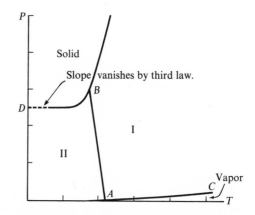

FIGURE 4.2
Phase diagram of liquid helium.

The transition I–II is not a phase change of the first kind since there is no latent heat or discontinuity in specific volume. Equations holding at a λ-transition have been given in Section 3.6.

The other transitions, including DB, are of the first kind. The slope of DB is therefore

$$\frac{dP}{dT} = \frac{S(\text{liquid}) - S(\text{solid})}{V(\text{liquid}) - V(\text{solid})} = \frac{\Delta S}{\Delta V}.$$

The slope of DB is positive but quickly flattens out below the λ-transition. Where DB is flat, the liquid has the same entropy as the solid and therefore it is as highly ordered, in contrast to the usual situation.

4.3 Helium II and Superfluidity

Helium II is characterized by anomalies in the thermal conductivity and viscosity. We shall also discuss the mechanocaloric effect and second sound.

Thermal conductivity

The heat flow is not proportional to ΔT.

Viscosity

The mass flow through a capillary is not proportional to ΔP.

If one tries to write the mass flow through a capillary tube in the form (Poiseuille) $\sim \eta_c^{-1} \Delta P$ where η_c is the capillary coefficient of viscosity, then $\eta_c^{-1} \cong \infty$. On the other hand, if a solid body moves through liquid helium II, the motion is damped by the normal viscosity of the liquid; i.e., the viscosity, η, determined in this second way, does not become very small below the λ-point.

Mechanocaloric Effect

A wall with capillary openings which permits helium II to escape is called a superleak. When liquid II escapes from a container through a superleak, the liquid remaining behind becomes warmer; when it flows into a container through a superleak the liquid already there is cooled.

The inverse effect is also seen, i.e., when the temperature is raised in a local region enclosed by a superleak the resulting temperature difference is removed by some kind of flow through the superleak. This flow results in an elevated pressure and may be exhibited by the fountain effect [5].

Second Sound

Ordinary heat conductivity is governed by the equation

$$\nabla^2 T = a \frac{\partial T}{\partial t}.$$

This equation is linear in t, and the distance that a temperature fluctuation progresses in the time t is $\sim \sqrt{t}$.

In helium II, however, heat conductivity is described by a wave equation

$$\nabla^2 T = \frac{1}{C^2} \frac{\partial^2 T}{\partial t^2}.$$

Hence distance is related to time linearly. The corresponding velocity is called the velocity of second sound.

4.4   Two Fluid Model

The properties of helium II can be correlated by supposing that it is a mixture of two fluids, a normal fluid and a superfluid with densities ρ_n and ρ_s respectively. Then

$$\rho = \rho_n + \rho_s. \tag{4.1}$$

We suppose that

$$\frac{d\rho_s}{dT} < 0 \qquad \text{below } \lambda\text{-point} \tag{4.1a}$$

and

$$\rho_s = 0 \qquad T \geq T_\lambda \tag{4.1b}$$

$$\rho_n = 0 \qquad T = 0. \tag{4.1c}$$

The superfluid is postulated to have no viscosity and to carry no entropy and to be entirely responsible for the unusual properties of helium II. The super-

FIGURE 4.3
Cooling by adding superfluid.

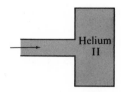

fluid behaves as if it were at absolute zero; it is therefore perfectly ordered, carries no entropy, and cannot interact with its environment unless it is excited (converted to or coupled to normal fluid). Although the concept of the superfluid is not a rigorous one, we shall now see how it helps us to understand the properties of helium II.

Viscosity

Since the superfluid has no viscosity it will escape through capillaries without resistance. Hence the viscosity of helium II, as measured by the capillary method, will be the same as the viscosity of the superfluid. On the other hand if the viscosity is determined by damping the motion of a solid body in the fluid, then only normal fluid will contribute to the damping and hence the total viscosity will, under these conditions, be proportional to the fraction of normal fluid.

Heat Conduction

Let an amount of superfluid flow into a thermally isolated sample of helium II (Figure 4.3). It will increase the mass but not the entropy. Hence the entropy per unit mass will be decreased. But

$$T \, dS = c \, dT. \qquad (4.2)$$

If $dS < 0$, we have $dT < 0$. Hence the introduction of superfluid will cool the helium II, and removal of superfluid will warm helium II.

If we now consider two volumes at temperatures T_A and T_B connected by a capillary (Figure 4.4), we see that the temperature difference can be removed by the flow of the superfluid. There is no entropy change associated with this flow. It would be described as reversible heat conduction ($\Delta S = 0$) but of course the mechanism is not at all the usual one for heat conduction.

Mechanocaloric and Fountain Effect

If the region A is warmed, the resulting temperature difference would normally disappear because of irreversible heat conduction. This process

FIGURE 4.4
Heat conduction by superfluid ($T_B > T_A$).

would be much slower than flow of superfluid and does not operate here. Instead there is a reversible heat flow.

Suppose now that the temperature difference $T_A - T_B$ is maintained by continuously supplying energy to A. Then superfluid will flow into A to set up the pressure difference $P_A - P_B$. Bearing in mind the approximate nature of the concept of superfluid, let us now try to relate $P_A - P_B$ to $T_A - T_B$.

Assume that the superfluid transports mass but neither energy nor momentum through the capillary [6]. Then the kind of equilibrium set up will be the same as if the fluid were separated into two regions by a rigid, adiabatic but permeable wall. Such a wall would not permit the transport of momentum or energy but would allow matter to pass through it. It will be shown in Chapter 5 that under these conditions the Gibbs function must be equal in the two parts:

$$g_A = g_B. \tag{4.3}$$

It follows that

$$\Delta g_A = \Delta g_B = 0 \tag{4.4}$$

if the pressure and temperature are unchanged in B. Then in A,

$$\frac{\Delta P}{\Delta T} = \left(\frac{\partial P}{\partial T}\right)_g = \frac{S}{V}. \tag{4.5}$$

The left side of this equation may be determined from a measurement of the pressure generated by the fountain effect while the right side may be calculated from the specific heat according to the formula:

$$S = \int c_P \frac{dT}{T}. \tag{4.6}$$

The preceding analysis may in this way be checked by independent measurements of the fountain effect and the specific heat.

Second Sound

In helium II a fluctuation in $\rho_s + \rho_n$ is propagated as a density wave or first sound. A simultaneous variation in $\rho_s - \rho_n$ induces the propagation of a temperature wave or second sound. If one pushes the model a little further one may obtain equations of the following structure [7]:

$$\left(\nabla^2 - \frac{1}{C_1^2}\frac{\partial^2}{\partial t^2}\right)\rho + \gamma_1 \nabla^2 T = 0 \tag{4.7a}$$

$$\left(\nabla^2 - \frac{1}{C_2^2}\frac{\partial^2}{\partial t^2}\right)T + \gamma_2 \frac{\partial^2 \rho}{\partial t^2} = 0 \tag{4.7b}$$

where $\rho = \rho_s + \rho_n$, and γ_1 and γ_2 are two functions that vanish as the temperature T approaches zero. Therefore at $T = 0$, the two waves decouple: C_1 and C_2 are known as the velocities of first and second sound, which are density and temperature waves, respectively.

It is also possible to excite sound waves exclusively in the superfluid component by preventing the normal fluid from moving. In this kind of experiment the helium II fills a porous solid matrix which permits only the superfluid to move. (The solid matrix is thus a superleak.) Under these circumstances only one of the two modes normally present is able to propagate. This mode is known as fourth sound [8] and its velocity is given by the approximate expression:

$$C_4 \cong \left(\frac{\rho_s}{\rho} C_1{}^2 + \frac{\rho_n}{\rho} C_2{}^2\right)^{1/2}. \tag{4.8}$$

Therefore at absolute zero C_4 approaches C_1 while at the λ-point C_4 approaches C_2.

The wavelike propagation of temperature is direct evidence of a reversible mechanism for heat transport; the ordinary mechanism, which is a diffusion of fast molecules according to the random walk and the law $x \sim \sqrt{t}$, never leads to wavelike propagation and a characteristic velocity.

4.5 Landau Criterion for Superfluidity

In order that a solid experience viscosity in moving through a fluid, there must be a mechanism for the loss of energy and momentum from the solid to the liquid. In addition it must be possible to satisfy the conservation laws of energy and momentum:

$$\frac{1}{2} MV^2 = \frac{1}{2} MV'^2 + \epsilon \tag{5.1}$$

$$MV = MV' + \mathbf{p} \tag{5.2}$$

where ϵ and $\mathbf{p}$ are the energy and momentum given to the fluid by the moving solid when its velocity changes from $\mathbf{V}$ to $\mathbf{V}'$. Then ϵ and $\mathbf{p}$ characterize some excitation of the fluid. Equations (5.1) and (5.2) may be rewritten as follows:

$$\frac{M}{2} (\mathbf{V} + \mathbf{V}')(\mathbf{V} - \mathbf{V}') = \epsilon$$

$$\frac{1}{2} (\mathbf{V} + \mathbf{V}')\mathbf{p} = \epsilon.$$

The maximum value of $\frac{1}{2} (\mathbf{V} + \mathbf{V}')\mathbf{p}$ is $\frac{1}{2} |\mathbf{V} + \mathbf{V}'|p$, which is less than Vp.

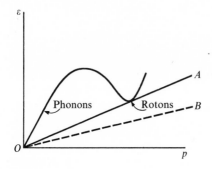

FIGURE 4.5
The energy spectrum of liquid helium II.

Hence

$$Vp > \epsilon$$

or

$$V > \epsilon/p \tag{5.3}$$

in order that the conservation laws be satisfied. This is the Landau criterion for superfluidity. The excitation curve [9] for the superfluid phase of helium (which has been determined by neutron diffraction and specific heat measurements) is shown in Figure 4.5. Here the energy of an excitation is plotted against its momentum. Let the velocity of the moving solid be represented by the slope of OB. If the slope of OB is less than the slope of the extreme tangent OA, equation (5.3) cannot be satisfied. Therefore such a slow moving solid cannot give up energy and momentum to the fluid without violating the conservation laws. Finally we remark that superfluidity should always appear in such a model if the tangent with least slope does not coincide with the p-axis.

4.6   Critical Velocity

The value of the critical velocity (at which viscosity ought to appear) is the slope of OA and comes out to be 60 m/sec—this velocity is two orders of magnitude greater than experiment indicates and does not depend on the width of the capillary, which is also in disagreement with observation. It follows that neither the phonons nor the rotons associated with Figure 4.5 are responsible for the onset of viscosity. In fact a different kind of experiment [10] has revealed that the excitations associated with the disappearance of superfluidity are quantized vortex rings with circulation

$$\oint \mathbf{V} \, d\mathbf{l} = nh/M \tag{6.1}$$

where M is the mass of the helium atom. The core of the vortex ring is of the

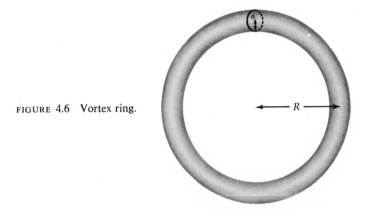

FIGURE 4.6   Vortex ring.

order of an atomic radius, and the radius of the ring is of the order of $10^2 \sim 10^4$ atoms.

When the Landau criterion for superfluidity is applied to this branch of the spectrum, one obtains by computing ϵ and p for a vortex ring [11]

$$V_{\text{critical}} = \frac{\hbar}{RM} \ln \frac{R}{a} \tag{6.2}$$

where R is the radius of the ring and a is the radius of the vortex core (Figure 4.6). This last formula gives values of the critical velocity that agree rather well with experiment, and in addition it correctly predicts that RV_{critical} is a constant, when R is the radius of the capillary.

According to the picture just described, the superfluid properties are to be ascribed to the scarcity of weak excitations or low lying states of helium II. When the superfluid is weakly excited, then those excitations that are most easily produced are vortex rings.

When the superfluid is "pure", there are no vortex excitations and therefore

$$\oint \mathbf{V} \, d\mathbf{l} = 0 \tag{6.3}$$

or

$$\text{curl } \mathbf{V} = 0. \tag{6.3a}$$

4.7   Superconductivity

Superconductors are classified into two groups, namely, type I or soft, and type II or hard, and differ rather widely among themselves [12]. We are here interested in their essential generic properties rather than in their individual differences and we shall therefore give a very simplified picture.

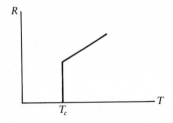

FIGURE 4.7
Resistivity of superconductor.

The situation is illustrated in Figure 4.7: below a certain transition temperature (T_c) the electrical resistance vanishes completely. The transition takes place over a temperature interval of the order of 10^{-2} degrees.

If a torus of superconducting material situated in a magnetic field is cooled below its transition temperature and the field is then switched off, a current is induced in the ring and continues to flow with undiminished strength; the damping of the current is too small to be measured. The lifetime appears to be greater than 10^5 years [13].

By increasing the magnetic field the superconducting properties may be destroyed. The locus of transition temperatures is shown in Figure 4.8 if the superconductor is soft or of type I [14].

Suppose we try to describe the superconducting state by Ohm's law

$$\mathbf{i} = \sigma \mathbf{E}$$

where

$$\sigma = \infty. \tag{7.1}$$

If $\mathbf{i}$ is finite, $\mathbf{E} = 0$ and curl $\mathbf{E} = 0$; but by the macroscopic Maxwell equation

$$\operatorname{curl} \mathbf{E} = -\frac{1}{c}\frac{\partial \mathbf{B}}{\partial t},$$

it follows that

$$\frac{\partial \mathbf{B}}{\partial t} = 0.$$

One would conclude from this argument that $\mathbf{B}$ could not change in a superconductor. It would then follow that the magnetic field in the superconductor

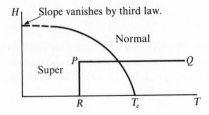

FIGURE 4.8
Transition curve dividing
normal and superconducting phases.

at point P would depend on whether P were approached from Q or from R (Figure 4.8). If P were approached from R, one would expect no field since there is none at R. On the other hand if P were approached from Q, one might expect the same field as at Q.

However, experiment (the Meissner effect) shows that the magnetic field is expelled from the interior of the ring at the same time that it becomes superconducting. The point P, if it is not too close to the surface of the superconductor, is characterized by the equation

$$\mathbf{B} = 0 \tag{7.2}$$

independently of how it is approached. The earlier argument to the contrary is incorrect because the constitutive equations do not hold in their usual form. In superconductors, the macroscopic Maxwell-Lorentz equations have to be modified; the new equations are known as Maxwell-London equations.

In the situation just described there is a so-called complete Meissner effect, that is, $\mathbf{B}$ vanishes everywhere in the interior of the superconductor except for a transitional surface layer, whose thickness ($\sim 500\ \mathring{A}$) is called the penetration depth.

4.8 Thermodynamics of Transition

As we have just seen, the superconducting state is a perfect diamagnetic, i.e., it is described by the equation

$$\mathbf{B}_s = \mathbf{H}_s + 4\pi \mathbf{M}_s = 0. \tag{8.1}$$

On the other hand, a normal conductor is only weakly paramagnetic. Hence

$$\mathbf{B}_n = \mathbf{H}_n + 4\pi \mathbf{M}_n$$
$$\mathbf{M}_n \cong 0 \tag{8.2}$$

in the normal phase. Thus there is a discontinuity in $\mathbf{M}$ of amount

$$\Delta \mathbf{M} = \mathbf{M}_s - \mathbf{M}_n = -\frac{\mathbf{H}_s}{4\pi}. \tag{8.3}$$

The first law is

$$dQ = dU - \mathbf{H}\,d\mathbf{M}. \tag{8.4}$$

The corresponding Gibbs function is

$$G = U - MH - TS \tag{8.5}$$

and the Clausius-Clapeyron equation becomes

$$\frac{dH}{dT} = -\frac{\Delta S}{\Delta M} \tag{8.6}$$

after the substitution $P \longrightarrow -H$ and $V \longrightarrow M$ in equations (4.3) and (5.6) of Chapter 3. Hence

$$-\frac{H}{4\pi}\frac{dH}{dT} = S_n - S_s. \tag{8.7}$$

Empirically, as Figure 4.8 shows,

$$\frac{dH}{dT} < 0. \tag{8.8}$$

Hence

$$S_s < S_n \tag{8.9}$$

or the superconducting state is more highly ordered than the normal state except at $H = 0$ where $S_n = S_s$. In addition one finds that the empirical slope dH/dT approaches zero as T approaches zero (see Figure 4.8). One therefore concludes that

$$\lim_{T \to 0} (S_n - S_s) = 0. \tag{8.10}$$

The situation here is exactly like that encountered in the transition between liquid and solid helium, and we shall see that these are both illustrations of the third law.

Let us now differentiate (8.7):

$$\frac{1}{4\pi}\frac{d}{dT}\left(H\frac{dH}{dT}\right) = \frac{dS_s}{dT} - \frac{dS_n}{dT}$$

$$\frac{1}{4\pi}\left[T\left(\frac{dH}{dT}\right)^2 + TH\frac{d^2H}{dT^2}\right] = c_s - c_n. \tag{8.11}$$

At $H = 0$,

$$\Delta c = \frac{1}{4\pi}T\left(\frac{dH}{dT}\right)^2. \tag{8.12}$$

Hence at $H = 0$ we have a discontinuity in specific heat although S and M are continuous, and therefore a phase transition of the second kind. This interpretation may be checked by comparing the measured slope of the transition curve with the discontinuity in the specific heat [15].

4.9   London Equations

In superconductors Maxwell's equations are of course still valid in their microscopic form. However, their macroscopic form becomes altered as follows:

$$\mathbf{j} = \mathbf{j}_s + \mathbf{j}_n \tag{9.1}$$

$$\mathbf{j}_n = \sigma \mathbf{E} \tag{9.2}$$

$$\frac{\partial}{\partial t} (\Lambda \mathbf{j}_s) = \mathbf{E} \tag{9.3}$$

$$\text{curl} (\Lambda \mathbf{j}_s) = -\frac{1}{c} \mathbf{H} \tag{9.4}$$

where Λ is a constant of the London theory [3] and determines the penetration thickness [16].

According to these equations the total current consists of a normal part ($\mathbf{j}_n$) plus a superconducting part ($\mathbf{j}_s$). The normal part obeys Ohm's law but the supercurrent is determined by equations (9.3) and (9.4). Therefore, a steady supercurrent is determined by the magnetic rather than the electric field. The new feature introduced by London may be expressed concisely by postulating that the vector potential is related to the supercurrent as follows:

$$\mathbf{A} = -c \, \Lambda \mathbf{j}_s. \tag{9.5}$$

These equations have solutions that correctly describe the persistent currents (steady currents in the absence of field) and also the perfect diamagnetism at low frequencies. One may summarize by saying that the observed electromagnetic properties can be correlated with the aid of the postulate (9.5). An intuitive argument for this relation is given in Appendix C.

4.10   Quantization of Magnetic Flux

The quantum mechanical character of the superconducting state is shown directly by the quantization of magnetic flux threading a superconducting ring. Let the flux be

$$\Phi = \int B_n \, dS. \tag{10.1}$$

Then it is observed that Φ may take on only the discrete values [17]

$$\Phi_n = n \left(\frac{\hbar c}{2e} \right). \tag{10.2}$$

This effect is analogous to the existence of quantized circulations in helium II, which appear in units of $\hbar/M$. The quantized flux and quantized vortex motion are clear examples of quantum phenomena, for which one has

$$pl \sim n\hbar \qquad (10.3)$$

where p is a characteristic momentum and l is a characteristic length. In atomic physics l is of the order of an atomic radius and p corresponds to an energy of the order of an electron volt. If l is to become macroscopic, p must become relatively very small. That is possible at low temperatures and a quantization condition of the above type is realized, with l macroscopic, for superfluids and superconductors.

4.11   Microscopic Theory

The electronic specific heat of a normal metal varies linearly with the temperature, but for a superconductor one usually has

$$c \sim e^{-T_0/T}. \qquad (11.1)$$

Such behavior is evidence for the existence of an effective gap [18] in the energy spectrum just above the ground state, i.e., a scarcity of low lying states. The B.C.S. atomic theory of superconductivity [19] is able to offer a mechanism for the appearance of such a gap and also to obtain the key London equation (9.5) which permits the phenomenological correlation of the electromagnetic properties of a superconductor.

According to the B.C.S. theory the peculiar nature of the ground state arises as a result of the attraction between pairs of electrons situated at the Fermi surface and having opposite momenta and spins. Direct evidence of the pairing is provided by the fact that the unit of magnetic flux is $\hbar c/2e$ rather than $\hbar c/e$. The attractive electronic interaction is mediated by the lattice; evidence for this remark is afforded not only by the sign of the interaction but also by the isotope effect, i.e., the dependence of the transition temperature on the mass of the lattice ion:

$$T_c \sim 1/\sqrt{M}. \qquad (11.2)$$

Finally it may be noted that the pairs in the ground state greatly overlap in space and that there are very strong pair-pair correlations in addition to correlation between mates of the same pair. The superconducting state is highly ordered in this sense.

4.12   General Remarks on Superconductivity and Superfluidity

Thermodynamics is concerned with general properties of matter rather than with specific mechanisms. From this viewpoint superconductivity and superfluidity are interesting because they illustrate the general trend of matter to become more highly organized as one approaches absolute zero. A physical system may be said to be completely organized when it is described by a single quantum mechanical state, but that situation is not reached at finite values of $1/T$. Any actual system is characterized by a set of states; the two fluid theories, which have been worked out for both superconductors and superfluids, correspond to the true picture in roughly the following way: the superfluid corresponds to the pure state which alone would be realized at absolute zero, while the normal fluid corresponds to the admixed states; and when the observed behavior is correlated with a mixture of normal fluid and superfluid, one is attempting to take into account the complete set of states.

The striking properties of these systems result from the way in which they are decoupled from their environments. That is, the superfluid has no viscosity because it cannot interact with other matter without violating the conservation laws; the same qualitative explanation accounts for the absence of resistance in superconductors. These systems are decoupled from their environments by quantum mechanics in essentially the same way as the hydrogen atom, a small molecule, or a biological molecule. All these systems are stabilized by the conservation laws in conjunction with the scarcity of interactions allowed by the quantum mechanical spectrum and matrix elements.

4.13   The Experimental Basis of the Third Law

We have seen two examples of the third law:

(a)
$$\lim_{T \to 0} \left(\frac{dP}{dT}\right) = 0, \qquad (13.1)$$

i.e., the slope of the phase boundary between solid and liquid helium II approaches zero as $T \longrightarrow 0$ (see Figure 4.2); and

(b)
$$\lim_{T \to 0} \left(\frac{dH}{dT}\right) = 0, \qquad (13.2)$$

i.e., the slope of the phase boundary between the superconducting and normal state approaches zero at absolute zero (see Figure 4.8).

In both cases the Clausius-Clapeyron relation implies

$$\lim_{T \to 0} \Delta S = 0. \tag{13.3}$$

The third law may be stated in the following form (Nernst-Simon): The entropy change associated with any isothermal reversible process in a condensed system approaches zero as the temperature approaches zero.

Formally we also have statements of the following kind:

$$\lim_{T \to 0} \left(\frac{\partial S}{\partial P}\right)_T = 0. \tag{13.4}$$

This statement implies, by one of Maxwell's relations,

$$\lim_{T \to 0} \frac{1}{V} \left(\frac{\partial V}{\partial T}\right)_P = 0. \tag{13.5}$$

Therefore, the thermal expansion coefficient must also approach zero.

The statement analogous to (13.4) for magnetic systems is

$$\lim_{T \to 0} \left(\frac{\partial S}{\partial H}\right)_T = 0, \tag{13.6}$$

which implies

$$\lim_{T \to 0} \left(\frac{\partial M}{\partial T}\right)_H = 0. \tag{13.7}$$

According to this last statement of the third law, Curie's law must always break down at low temperatures.

In general the third law may be expressed in either of two ways which must be equivalent by virtue of the second law. That is, one may write

$$\lim_{T \to 0} \left(\frac{\partial \beta}{\partial T}\right)_S = 0 \tag{13.8}$$

or

$$\lim_{T \to 0} \left(\frac{\partial S}{\partial \alpha}\right)_T = 0 \tag{13.9}$$

where $\alpha d\beta$ is the corresponding work term. The first of these equations requires $\beta(T)$ to approach zero with a flat tangent. The second equation requires isothermals and isentropics to approach coincidence as $T \longrightarrow 0$.

By (2.21) of Chapter 3, this last statement implies

$$\lim_{T \to 0} \frac{C_P}{C_V} = 1.$$

Of course, since $C\,dT = T\,dS$, it is also true that $C_P \longrightarrow 0$, and $C_V \longrightarrow 0$ separately.

4.14   Unattainability Formulation of Third Law

Suppose that it were possible to reach $T = 0$ by some reversible path. Let the state so reached be f (Figure 4.9). Then f may be connected with the initial state by a reversible path consisting of segments of isothermals and isentropics. Let the last step be the isentropic nf. Then

$$S_n = S_f. \tag{14.1}$$

We also have the relation

$$S_n - S_a = \int_0^{T_n} c_a \frac{dT}{T} > 0. \tag{14.2}$$

It follows from (14.1) and (14.2) that

$$S_f - S_a > 0. \tag{14.3}$$

Therefore if it were possible to reach $T = 0$ we would have a violation of the Nernst formulation of the third law. An interesting illustration of the difficulty of reaching very low temperatures is provided by the example of adiabatic demagnetization mentioned in the previous chapter.

The statement of the third law in the form that absolute zero is not attainable in a finite number of steps is similar to the relativistic statement that the velocity of light is not attainable. On the velocity scale, high energy physics lies in the interval

$$\epsilon > 1 - \frac{V}{c} > 0$$

or

$$\frac{1}{\epsilon} < \frac{1}{1 - \frac{V}{c}} < \infty.$$

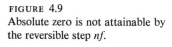

FIGURE 4.9
Absolute zero is not attainable by
the reversible step nf.

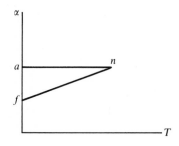

Similarly low energy physics lies in the interval

$$\epsilon > T > 0$$

or

$$\frac{1}{\epsilon} < \frac{1}{T} < \infty.$$

In both cases the limits are not attainable and the experimental domain, embracing all of high and low energy technology, is unbounded.

The unattainability statement may be paraphrased by remarking that at absolute zero a macroscopic system would be described by a single quantum mechanical state. The unattainability statement therefore implies that a macroscopic system cannot be reduced to a single quantum mechanical state. Nevertheless as absolute zero is approached, such a state will be simulated, as one sees in the examples of superconductors and superfluids.

All of the remarks in this section are based on the Nernst induction from those experiments that show that the isentropic surfaces approach coincidence with the isothermal surfaces as the temperature approaches absolute zero.

Notes and References

1. M. Zemansky, *Heat and Thermodynamics* (McGraw-Hill, New York, 1959).
2. W. H. Keesom, *Helium* (Elsevier, New York, 1942).
3. F. London, *Superfluids*, Vol. I (Wiley, New York, 1950).
4. Fairbank has shown that to within 10^{-6} degrees of the lambda point the specific heat behaves as if it were approaching a logarithmic infinity. M. J. Buckingham and W. M. Fairbank, *Progress in Low-Temperature Physics*, Vol. 3, p. 85 (North-Holland, Amsterdam, 1961).
5. If light is allowed to fall on a bulb of helium II containing black powder which absorbs the radiant energy, the temperature of the surrounding fluid is slightly increased. If this bulb is connected by a small opening with a surrounding bath of liquid II, there will be a flow into the bulb which produces a marked increase in pressure or a fountain.
6. We may just as well assume that the volumes A and B are separated not by a single capillary but by a rigid adiabatic wall which is also a super-leak.
7. K. Huang, *Statistical Mechanics* (Wiley, New York, 1963). See p. 405.
8. The theoretical expression for fourth sound agrees rather well with the

experiments of K. A. Shapiro and I. Rudnick, *Phys. Rev.* 137, A1383 (1965).

"Third sound" is reserved for waves in a thin film of helium II resulting from body forces. Here also the normal fluid does not move.

9. The slope of the linear portion of $\epsilon(p)$ is the velocity of sound, c, i.e.,

$$\epsilon = cp.$$

The corresponding excitations are called phonons.

Near the minimum p_0 in the curve we have

$$\epsilon = \Delta + \frac{(p - p_0)^2}{2\mu}$$

where

$$\Delta/k = 8.6°\text{K}$$

$$p_0/\hbar = 1.91 \ A^{-1}$$

$$\mu = 0.16 \ M(\text{He}).$$

The excitations in this part of the spectrum are called rotons.

10. G. W. Rayfield and F. Reif, *Phys. Rev. Letters* 11: 305 (1963).

11. I. M. Khalatnikov, *Introduction to the Theory of Superfluidity*, p. 99 (Benjamin, New York, 1965).

12. F. London, *Superfluids*, Vol. II (Wiley, New York, 1950). See also de Gennes, reference [14] and Schrieffer, reference [19].

13. File and Mills, *Phys. Rev. Letters* 10: 93 (1963).

14. The corresponding plot for type II or hard superconductors is shown in Figure 4.10. In regions (a), (b), and (c) there is complete, partial, and no Meissner effect respectively. In (c) there is only surface superconductivity. See P. G. de Gennes, *Superconductivity of Metals and Alloys* (Benjamin, New York, 1966).

15. See, for example, reference [1].

16. The penetration depth is

$$\lambda = (\Lambda c^2/4\pi)^{1/2}.$$

Then

$$B(x) = B(0)e^{-x/\lambda}.$$

17. Deaver and Fairbank, *Phys. Rev. Letters* 7: 43 (1961). Since $\mathbf{B} = 0$

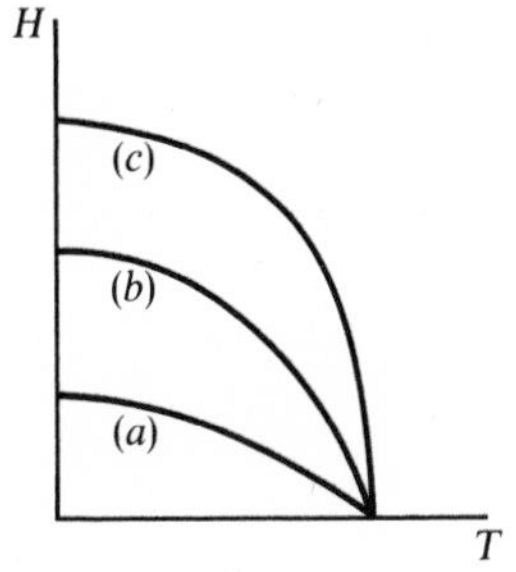

FIGURE 4.10
Type II superconductors. In regions (a), (b) and (c) respectively, there is complete, partial, and no Meissner effect.

inside the superconductor, the contour in (10.1) may be taken anywhere in the ring. Conversely the argument may be turned around to derive the Meissner effect.

18. The energy gap may also be determined (a) as the threshold for absorbing electromagnetic radiation and (b) as a threshold for tunneling current between two films of superconducting material separated by a thin oxide layer. See reference [19].

19. A brief sketch of this theory is given here in Chapter 10. For a more complete discussion, see, for example, J. R. Schrieffer, *Theory of Superconductivity* (Benjamin, New York, 1964).

5

General Conditions of Thermodynamic Equilibrium and Other Applications

5.1 Second Law for Open Systems Subject to General Interactions

The discussion so far has been limited to systems of given mass with only two degrees of freedom except when it was necessary as a matter of principle to consider more general systems. Using the methods introduced by Gibbs [1] we shall now briefly illustrate the power of thermodynamics to make important statements about very complicated physical systems.

The second law has been written in the following form for reversible changes:

$$dU = T \, dS - P \, dV. \tag{1.1}$$

In the preceding equation it has been assumed that the physical system is homogeneous with temperature T and pressure P and has a given mass, say M. Then the internal energy U is a function of S, V, and M:

$$U = U(S, V, M). \tag{1.2}$$

A system with given mass M is called *closed*. We consider now *open* systems, namely systems for which M is not fixed. Then

$$dU = T \, dS - P \, dV + \left(\frac{\partial U}{\partial M}\right)_{SV} dM. \tag{1.3}$$

The system may in fact contain several components each of which can be exchanged with a reservoir. If the interaction with the environment is also general, one has

$$dU = T\,dS - \sum_r P_r\,dV_r + \sum_a \left(\frac{\partial U}{\partial M^a}\right)_{SV} dM^a \tag{1.4}$$

where P_r and V_r are generalized pressures and volumes corresponding to different kinds of work, and where M^a is the mass of the ath component.

Then the Gibbs function is

$$G = U - TS + \sum_r P_r V_r \tag{1.5}$$

and

$$dG = -S\,dT + \sum_r V_r\,dP_r + \sum_a \left(\frac{\partial U}{\partial M^a}\right)_{SV} dM^a. \tag{1.6}$$

By the formula for partial differentiation,

$$dG = \left(\frac{\partial G}{\partial T}\right)_{PM} dT + \sum_r \left(\frac{\partial G}{\partial P_r}\right)_{TM} dP_r + \sum_a \left(\frac{\partial G}{\partial M^a}\right)_{TP} dM^a \tag{1.7}$$

so that

$$S = -\left(\frac{\partial G}{\partial T}\right)_{PM} \qquad V_r = \left(\frac{\partial G}{\partial P_r}\right)_{TM} \tag{1.8}$$

and

$$\left(\frac{\partial G}{\partial M^a}\right)_{TP} = \left(\frac{\partial U}{\partial M^a}\right)_{SV}. \tag{1.9}$$

These relations are subject to compatibility or reciprocity conditions such as the following:

$$\left(\frac{\partial V_r}{\partial P_s}\right)_{TM} = \left(\frac{\partial V_s}{\partial P_r}\right)_{TM}, \tag{1.10}$$

similar to relations that appeared earlier in Chapter 2, equation (4.9).

The systems considered in this section can exchange mass, as well as energy and momentum with their environment. This transport may go on internally as well, and the open system may be regarded as a subsystem of a closed larger system. A system exchanging mass, energy, and momentum with its surroundings is described thermodynamically as an open system that is doing nonadiabatic work.

5.2 Chemical Reactions

The composition of the system may change not only as a result of mass flow through the boundary but also as a result of internal reactions. At lower temperatures these reactions are molecular and at higher temperatures they are nuclear. The thermodynamic description is formally the same in both cases.

We shall now assume that the total system is closed and that the temperature and generalized pressures are controlled, so that the slightly generalized equation (4.7) of Chapter 3 is appropriate:

$$dG \leq -S\,dT + \sum_r V_r\,dP_r.$$

If the temperature and pressures are held fixed,

$$dG \leq 0. \tag{2.1}$$

In irreversible changes G decreases, and the condition for equilibrium is that G can no longer decrease, or that the Gibbs function is a minimum, subject of course to any additional constraints. In this case the constraints express the conservation of mass; and since they are determined by the chemical reactions, it is natural to specify the system by mol numbers $N_1, \ldots, N_c$ instead of mass numbers $M_1, \ldots, M_c$. Equation (2.1) then implies

$$\sum_j \frac{\partial G}{\partial N_j} \delta N_j = 0 \tag{2.1a}$$

where δN_j means a virtual variation and where the conservation of mass is expressed by equations of the following type:

$$\delta N_j = \nu_j\,\delta\lambda. \tag{2.2}$$

For example, consider

$$2H_2O \longrightarrow 2H_2 + O_2,$$

then

$$\delta N(H_2O) = -2\delta\lambda$$
$$\delta N(H_2) = 2\delta\lambda$$
$$\delta N(O_2) = \delta\lambda.$$

An example of a nuclear reaction is

$$D + D \longrightarrow T + H$$

$$\delta N(D) = -2\delta\lambda$$

$$\delta N(T) = \delta\lambda$$

$$\delta N(H) = \delta\lambda.$$

If $\delta\lambda > 0$, the reaction is going from left to right. The ν_j in (2.2) mean the coefficients appearing in the chemical equation. The two conditions (2.1a) and (2.2) may be combined to give

$$\sum_k \nu_k \, \mu_k = 0 \tag{2.3}$$

where

$$\mu_k = \frac{\partial G}{\partial N_k}. \tag{2.4}$$

Equation (2.3) is the condition expressing chemical equilibrium. The μ_k are called chemical potentials. If several reactions are possible, the equilibrium is possible only if this equation is satisfied for each of the possible reactions separately.

5.3   Ideal Gases

Equation (2.3) is exact and to apply it one has in principle simply to calculate the μ_k for a gaseous mixture. To get some simple insight into the implications of this condition, consider a mixture of ideal gases. Strictly speaking, of course, ideal gases do not interact at all, but we are interested in conditions that hold only at equilibrium and not in the length of time required to reach equilibrium. It is therefore legitimate to consider ideal gas behavior; more accurate equations of state will of course lead to certain corrections to our equilibrium conditions.

For a mixture of noninteracting gases we have

$$U = \sum_i U_i \tag{3.1a}$$

$$P = \sum_i P_i \tag{3.1b}$$

$$S = \sum_i S_i. \tag{3.1c}$$

The additive property of the energy and the entropy [2] for noninteracting systems has already been assumed in Chapter 2. Here the noninteracting gases occupy the same volume and therefore the pressure or energy density is also additive.

Hence

$$G = \sum_i G_i. \tag{3.2}$$

In general,

$$\left(\frac{\partial G}{\partial P}\right)_T = V. \tag{3.3}$$

For an ideal gas,

$$\left(\frac{\partial G}{\partial P}\right)_T = \frac{nRT}{P} \tag{3.4}$$

with the integral

$$G = nRT \left[\ln P + \phi(T)\right]. \tag{3.5}$$

Hence for the mixture of perfect gases we have

$$\mu_k = RT \left[\ln P_k + \phi_k(T)\right]. \tag{3.5a}$$

By equation (2.3)

$$\sum_k \nu_k \left[\ln P_k + \phi_k(T)\right] = 0 \tag{3.6}$$

or

$$\prod_k P_k^{\nu_k} = K(T) \tag{3.6a}$$

where

$$\ln K(T) = - \sum_k \nu_k \, \phi_k(T). \tag{3.6b}$$

Equations (3.1a, b, c) express the equilibrium conditions (2.3) when the system is a simple mixture of ideal gases. We have, for example,

$$\frac{P(H_2O)^2}{P(H_2)^2 \, P(O_2)} = K(T)$$

and

$$\frac{P(D)^2}{P(T) \, P(H)} = K(T).$$

Notice that $K(T)$ depends on the temperature only. The two preceding equations are illustrations of the law of mass action. From the kinetic point of view the law of mass action expresses the condition that collisions between molecules are taking place in such a way that the rate from reactants to products equals the rate of the inverse reaction. The temperature dependence of K is already fixed by (3.6b). It is also interesting to write (3.6b) as a differential equation. One has by differentiating (3.6b)

$$\frac{d}{dT}(\ln K) = - \sum_k \nu_k \left(\frac{d\phi_k}{dT}\right) \tag{3.7}$$

and by (3.5a)

$$\frac{d\phi_k}{dT} = \left(\frac{\partial}{\partial T}\frac{\mu_k}{RT}\right)_P. \tag{3.7a}$$

It is possible to express the right side of (3.7a) simply in terms of heat of the reaction, by noting

$$\left(\frac{\partial}{\partial T}\frac{G}{RT}\right)_P = \frac{1}{RT}\left(\frac{\partial G}{\partial T}\right)_P - \frac{G}{RT^2}$$

$$= -\frac{H}{RT^2} \tag{3.8}$$

where H is the enthalpy.

Hence (3.7) becomes

$$\frac{d}{dT}(\ln K) = \frac{\displaystyle\sum_k \nu_k h_k}{RT^2} \tag{3.9}$$

or

$$\frac{d}{dT}(\ln K) = \frac{\Delta H}{RT^2} \tag{3.10}$$

where

$$\Delta H = \sum_k \nu_k h_k. \tag{3.10a}$$

Quite generally,

$$\Delta Q = \Delta H - V \, \Delta P. \tag{3.11}$$

Hence if $\Delta P = 0$,

$$\Delta Q_P = \Delta H. \tag{3.12}$$

Then (3.10) may be written

$$\frac{d}{dT}(\ln K) = \frac{Q_P}{RT^2} \tag{3.13}$$

where Q_P is the heat of reaction at constant pressure:

$$Q_P = \sum_k \nu_k h_k. \tag{3.13a}$$

The heat of reaction defined this way is referred to the numbers ν_k. Equation (3.13) is the van't Hoff equation.

If Q_P is approximately independent of T,

$$\ln K = -\frac{Q_P}{RT} + A. \tag{3.14}$$

Hence the plot of $\ln K$ against $1/T$ is, in such a case, a straight line whose slope gives Q_P/R. In the chemical case, Q_P is of the order of electron volts, while in the nuclear case it is of the order of 10^6 greater.

5.4   Equilibrium of Heterogeneous System

A physical system that is thermodynamically homogeneous is called a single phase. We now consider a system composed of several distinct phases. Under conditions of constant temperature and pressure, the equilibrium condition is

$$\delta G = 0. \tag{4.1}$$

Let the mass of the ath component in the α-phase be $M_\alpha{}^a$. Then the preceding condition also reads

$$\sum_{a=1}^{c} \sum_{\alpha=1}^{\phi} \hat{\mu}_\alpha{}^a \, \delta M_\alpha{}^a = 0 \tag{4.2}$$

where the sum is over c-components and ϕ-phases, and the coefficients

$$\hat{\mu}_\alpha{}^a = \frac{\partial G}{\partial M_\alpha{}^a} \tag{4.2a}$$

are specific instead of molal potentials. There are usually additional equations that have to be satisfied; these tell which chemical reactions, if any, may occur. Suppose there are no chemical reactions. Then

$$\sum_{\alpha=1}^{\phi} \delta M_\alpha{}^a = 0 \qquad a = 1, \ldots, c. \tag{4.3}$$

There are $c\phi$ variables and c conditions, and therefore $c\phi - c$ independent variables. By undetermined multipliers,

$$\sum_{a=1}^{c} \sum_{\alpha=1}^{\phi} (\hat{\mu}_\alpha{}^a - \lambda^a)\, \delta M_\alpha{}^a = 0. \tag{4.4}$$

Choose $\lambda^a = \hat{\mu}_1{}^a$; then only $c\phi - c$ terms remain. The corresponding $\delta M_\alpha{}^a$ are independent. Hence

$$\hat{\mu}_\alpha{}^a = \lambda^a \tag{4.5}$$

for all α.

If reactions are possible, the preceding argument is easily modified. One has only to interpret the $\delta M_\alpha{}^a$ as independent variables, i.e., quantities which may be varied independently. The number (c') of independent variables differs from the total number (c) of components by the number of possible reactions (r):

$$c' = c - r. \tag{4.6}$$

The equilibrium conditions read

$$\hat{\mu}_1{}^a = \hat{\mu}_2{}^a = \cdots = \hat{\mu}_\phi{}^a \qquad a = 1, \ldots, c' \tag{4.7}$$

i.e., the potentials of each independent component must be the same in all phases [3].

5.5   Remarks on Extensive and Intensive (Specific) Thermodynamic Variables

A homogeneous function of the nth degree is defined by the equation

$$F(\lambda x_1, \ldots) = \lambda^n F(x_1, \ldots). \tag{5.1}$$

It satisfies Euler's equation:

$$nF(x_1, \ldots) = \sum_k x_k \frac{\partial F}{\partial x_k}. \tag{5.2}$$

These relations are equivalent; either may be taken as a definition of a homogeneous function of nth degree.

By differentiation,

$$(n - 1)\frac{\partial F}{\partial x_j} = \sum_k x_k \frac{\partial}{\partial x_k}\left(\frac{\partial F}{\partial x_j}\right). \tag{5.3}$$

Hence $\partial F/\partial x_j$ is homogeneous of degree $n - 1$, if F is homogeneous of degree n.

With neglect of surface energies, the Gibbs function of a single phase is homogeneous of the first degree:

$$G(\lambda N_1, \ldots) = \lambda G(N_1, \ldots). \tag{5.4}$$

On the other hand,

$$\mu_k = \frac{\partial G}{\partial N_k} \tag{5.5}$$

is of zero degree:

$$\mu_k(\lambda N_1, \ldots) = \mu_k(N_1, \ldots). \tag{5.6}$$

Quantities of the first and zero degree are also called extensive and intensive (specific) respectively.

Choose $\lambda = N^{-1}$. Let $x_k = N_k/N$ (mol fractions). Then

$$\mu_k(x_1, \ldots) = \mu_k(N_1, \ldots), \tag{5.7}$$

but

$$\sum_k x_k = 1. \tag{5.8}$$

Hence the chemical potentials depend on only $c - 1$ independent variables. In just the same way the μ_k depend on only $c - 1$ mass fractions, $y_k = M_k/M$.

5.6   Phase Rule

There are $c'(\phi - 1)$ equilibrium conditions [equation (4.7)]

$$\mu_1{}^a = \cdots = \mu_\phi{}^a \qquad a = 1, \ldots, c' \tag{6.1}$$

connecting T, P, and the independent variables of composition. There are $\phi(c' - 1) + 2$ such variables which are independent by (5.8). Hence the number of degrees of freedom is

$$f = [\phi(c' - 1) + 2] - c'(\phi - 1) = c' - \phi + 2. \tag{6.2}$$

This is the Gibbs phase rule.

Examples

(a) Nonreacting mixture of c' gases. Here

$$c' = c, \qquad \phi = 1, \qquad f = c' - \phi + 2 = c + 1.$$

The independent variables may be identified as T, P, and the $c - 1$ mol fractions.

(b) Reacting mixture of gases:

$$2H_2O \rightleftharpoons 2H_2 + O_2.$$

Here $c' = c - 1 = 2$, $\phi = 1$, hence $f = 3$. The independent variables may be chosen as T, P, and $x(H_2)$. Then $x(H_2O)$ and $x(O_2)$ may be determined from

$$\frac{x(H_2O)^2}{x(H_2)^2\, x(O_2)} = K(T)\, P$$

and

$$\sum_i x_i = 1.$$

(c) Single component:

$$f = 3 - \phi.$$

The maximum number of phases that may coexist is three. Under conditions of very high pressure, many new phase transitions take place. There are known, for example, several phases of ice. The phase rule tells us, however, that one should never expect to find a quadruple point, no matter how complicated the phase diagram becomes.

(d) Two component system:

$$f = 4 - \phi.$$

In such a system we may have a quadruple point; this is called the eutectic point. A familiar example is the equilibrium of salt solution, salt, ice, and vapor.

5.7 General Conditions of Equilibrium and the Internal Energy

Instead of expressing the second law in terms of the Gibbs function, one may discuss the internal energy. Then

$$T\, \Delta S \geq \Delta U + \Delta W. \tag{7.1}$$

If we consider a mechanically isolated system, then

$$\Delta W = 0 \tag{7.2}$$

and therefore

$$T\, \Delta S \geq \Delta U. \tag{7.3}$$

Hence if (a) the internal energy is held fixed, the entropy tends to increase;

or if (b) the entropy is fixed, the internal energy tends to a minimum. In either case equilibrium is expressed by the two equations

$$\delta S = 0 \tag{7.4}$$

$$\delta U = 0. \tag{7.5}$$

In (a), the first equation expresses the fact that S has reached a maximum subject to a constraint on the internal energy. In (b), the second equation expresses the fact that U has reached a minimum while S is fixed. In both cases, the condition of mechanical isolation can be realized by enclosing the total system in a rigid wall. Then

$$\delta V = 0. \tag{7.6}$$

Let the various homogeneous regions or phases be separated by walls which may or may not permit the transport of energy, momentum, and mass. Then

$$\delta U = \Sigma_\alpha \, \delta U_\alpha = 0 \tag{7.7}$$

$$\delta S = \Sigma_\alpha \, \delta S_\alpha = 0 \tag{7.8}$$

$$\delta V = \Sigma_\alpha \, \delta V_\alpha = 0, \tag{7.9}$$

and   also

$$\delta M^a = \Sigma_\alpha \, \delta M_\alpha{}^a = 0 \qquad a = 1, \ldots, c \tag{7.10}$$

if the inner walls permit the transfer of mass between phases. There is one conservation equation for each of the c-components. We have

$$\delta U_\alpha = T_\alpha \, \delta S_\alpha - P_\alpha \, \delta V_\alpha + \Sigma_a \, \hat{\mu}_\alpha{}^a \, \delta M_\alpha{}^a. \tag{7.11}$$

By the method of Lagrangian multipliers,

$$\Sigma_\alpha(T_\alpha - T)\delta S_\alpha + \Sigma_\alpha(P - P_\alpha)\delta V_\alpha + \Sigma_{a\alpha}(\hat{\mu}_\alpha{}^a - \hat{\mu}^a)\delta M_\alpha{}^a = 0 \tag{7.12}$$

where T, P, and $\hat{\mu}^a$ are the multipliers.

By considering rigid, impermeable, but conducting walls, one finds only the conditions for temperature equilibrium. Then $\delta V_\alpha = \delta M_\alpha{}^a = 0$ and therefore

$$T_\alpha = T. \tag{7.13}$$

By considering adiabatic, impermeable, but flexible or movable walls, one finds the conditions for pressure equilibrium:

$$P_\alpha = P. \tag{7.14}$$

Finally if the walls are permeable but rigid and adiabatic, one gets

$$\hat{\mu}_\alpha{}^a = \hat{\mu}^a. \tag{7.15}$$

This result has been quoted in equation (4.3) of Chapter 4 in the discussion of superfluidity. Since the superfluid transports mass but neither momentum nor energy under the conditions discussed there, it follows that (7.15) is satisfied but not (7.14) and (7.13).

If the walls permit the transport of mass, momentum, and energy simultaneously then all three kinds of equilibrium are set up at the same time.

5.8   Stability and Thermodynamic Inequalities

Our discussion of thermodynamic equilibrium is so far incomplete. The condition already stated does not distinguish between maxima and minima. We have, for virtual changes in S and V,

$$\delta U = \left(\frac{\partial U}{\partial S}\right)\delta S + \left(\frac{\partial U}{\partial V}\right)\delta V + \frac{1}{2}\left[\frac{\partial^2 U}{\partial S^2}(\delta S)^2 + 2\frac{\partial^2 U}{\partial S\partial V}(\delta S)(\delta U)\right.$$
$$\left. + \left(\frac{\partial^2 U}{\partial V^2}\right)(\delta V)^2\right] \quad (8.1)$$

$$= T\,\delta S - P\,\delta V + \frac{1}{2}[\cdots]. \quad (8.1a)$$

Consider a single phase divided into two parts, A and B, by a conducting deformable wall (Figure 5.1). Let the total system be enclosed in an adiabatic rigid wall. Let us consider virtual changes in A and B such that

$$\delta S_A + \delta S_B = 0 \quad (8.2)$$
$$\delta V_A + \delta V_B = 0. \quad (8.3)$$

Since $T_A = T_B$ and $P_A = P_B$, the first order part of $\delta U_A + \delta U_B$ cancels, and the contributions of A and B to the second order are equal. It follows from (8.1) that

$$\delta U = \delta U_A + \delta U_B = U_{SS}(\delta S_A)^2 + 2U_{SV}(\delta S_A)(\delta V_A) + U_{VV}(\delta V_A)^2. \quad (8.4)$$

The stability condition is that

$$\delta U > 0. \quad (8.5)$$

Such a quadratic form is always positive only if

$$U_{SS} > 0, \quad \begin{vmatrix} U_{SS} & U_{SV} \\ U_{VS} & U_{VV} \end{vmatrix} > 0, \quad U_{VV} > 0. \quad (8.6)$$

FIGURE 5.1
Stable phase divided into two parts by a conducting, deformable wall.

Here

$$U_S = \left(\frac{\partial U}{\partial S}\right)_V = T \tag{8.7a}$$

and

$$U_{SS} = \left(\frac{\partial T}{\partial S}\right)_V. \tag{8.7b}$$

Hence $\qquad U_{SS} > 0$ implies $(\partial T/\partial S)_V > 0,$ $\qquad\qquad$ (8.8)

and

$$T\left(\frac{\partial S}{\partial T}\right)_V > 0$$

or

$$C_V > 0. \tag{8.9}$$

The other derivatives are

$$U_V = \left(\frac{\partial U}{\partial V}\right)_S = -P \tag{8.10a}$$

$$U_{VV} = -\left(\frac{\partial P}{\partial V}\right)_S. \tag{8.10b}$$

Hence $U_{VV} > 0$ implies

$$\left(\frac{\partial P}{\partial V}\right)_S > 0 \tag{8.11}$$

or that the isentropic compressibility, namely,

$$K_S = -\frac{1}{V}\left(\frac{\partial V}{\partial P}\right)_S,$$

is positive:

$$K_S > 0. \tag{8.12}$$

The condition on the determinant may be written

$$TV\beta_S^2 > C_V K_S \tag{8.13}$$

where β_S and K_S are isentropic thermal expansion and compressibility coefficients.

If the partition separating A from B permits the transport of matter, as well as energy and momentum, the stability condition requires that the following quadratic form be positive definite:

$$\delta U = \frac{1}{2} [U_{SS}(\delta S)^2 + U_{VV}(\delta V)^2 + U_{MM}(\delta M)^2 + 2U_{SV}(\delta S)(\delta V)$$
$$+ 2U_{SM}(\delta S)(\delta M) + 2U_{VM}(\delta V)(\delta M)]. \quad (8.14)$$

In other words the following determinants must be positive:

$$\begin{vmatrix} U_{SS} & U_{SV} & U_{SM} \\ U_{VS} & U_{VV} & U_{VM} \\ U_{MS} & U_{MV} & U_{MM} \end{vmatrix} > 0 \qquad \begin{vmatrix} U_{SS} & U_{SM} \\ U_{MS} & U_{MM} \end{vmatrix} > 0$$

$$\begin{vmatrix} U_{VV} & U_{VM} \\ U_{MV} & U_{MM} \end{vmatrix} > 0 \qquad\qquad U_{MM} > 0 \qquad (8.15)$$

as well as the conditions (8.6). For example, $U_{MM} = (\partial \hat{\mu}/\partial M)_{SV} > 0$ is analogous to the condition (8.9) that the specific heat be positive. If there is a fluctuation resulting in the instantaneous condition $T_A < T_B$, then the second law will demand heat flow from B to A to increase the entropy of the total system. However, equilibrium will not be restored by this flow unless the specific heat is positive. Similarly the second law demands the flow of matter from B to A if $\hat{\mu}_A < \hat{\mu}_B$, but again the result will not be equilibrium unless $\partial \hat{\mu}/\partial M > 0$.

It follows from the preceding analysis that systems that do not satisfy the given inequalities, such as systems with negative specific heat and negative compressibilities, cannot exist in stable thermodynamic equilibrium. Conversely, if a system is observed in stable thermodynamic equilibrium, it follows that these inequalities do hold for it.

Corresponding inequalities may be derived by considering other characteristic functions. However, it is important that the independent variables be extensive. It is then possible to consider fluctuations in the subsystems that occur in such a way that the sum of all the fluctuations vanishes. For example we may consider the second variation in A:

$$\delta^2 A = A_{VV}(\delta V)^2 + 2A_{VT}(\delta V)(\delta T) + A_{TT}(\delta T)^2. \quad (8.16)$$

Considering now the extensive variable V, one finds

$$A_{VV} = - \left(\frac{\partial P}{\partial V} \right)_T > 0. \quad (8.17)$$

Hence by equation (2.16), Chapter 3,

$$C_P - C_V = \frac{TV\beta^2}{K_T} \geq 0 \quad (8.18)$$

and by (2.21) of the same chapter

$$\frac{K_T}{K_S} = \frac{C_P}{C_V} \geq 1 \quad (8.19)$$

as was mentioned earlier.

5.9   Displacement of Chemical Equilibrium

When an equilibrium is stable, it can be displaced by altering the external constraints, for example, the temperature and pressure.

Consider a mixture of reacting gases at equilibrium. Let the pressure and temperature be changed. Then the reaction will run one way or the other until a new equilibrium is reached. Let the degree of reaction be ϵ. Then one has, for both the old and the new equilibrium,

$$\left(\frac{\partial G}{\partial \epsilon}\right)_{TP} = 0. \tag{9.1}$$

Hence

$$\Delta \left(\frac{\partial G}{\partial \epsilon}\right) = 0 \tag{9.2}$$

and

$$\frac{\partial^2 G}{\partial \epsilon\, \partial T} \Delta T + \frac{\partial^2 G}{\partial \epsilon\, \partial P} \Delta P + \frac{\partial^2 G}{\partial \epsilon^2} \Delta \epsilon = 0 \tag{9.2a}$$

where ΔT, ΔP are real (not virtual) changes in temperature and pressure, and where $\Delta \epsilon$ corresponds to the consequent shift of equilibrium. But

$$\frac{\partial G}{\partial T} = -S$$

$$\frac{\partial G}{\partial P} = V,$$

so that

$$-\frac{\partial S}{\partial \epsilon} \Delta T + \frac{\partial V}{\partial \epsilon} \Delta P + \frac{\partial^2 G}{\partial \epsilon^2} \Delta \epsilon = 0. \tag{9.3}$$

(a) Let $\Delta P = 0$. Then

$$\Delta \epsilon = \frac{(\partial S/\partial \epsilon)_{TP}\, \Delta T}{(\partial^2 G/\partial \epsilon^2)_{TP}} = \frac{1}{T} \frac{(\partial Q/\partial \epsilon)_{TP}\, \Delta T}{(\partial^2 G/\partial \epsilon^2)_{TP}}. \tag{9.4}$$

Since the Gibbs function is a minimum, one always has

$$\left(\frac{\partial^2 G}{\partial \epsilon^2}\right)_{TP} > 0. \tag{9.5}$$

Suppose for definiteness that the reaction is endothermic. Then

$$\left(\frac{\partial Q}{\partial \epsilon}\right)_{TP} > 0$$

and $\Delta T > 0$ implies $\Delta \epsilon > 0$ by (9.4). Hence increasing the temperature of an endothermic reaction causes the reaction to run in the direction that absorbs heat (lowering the temperature).

(b) Let $\Delta T = 0$. Then

$$\Delta \epsilon = -\frac{(\partial V/\partial \epsilon)\,\Delta P}{(\partial^2 G/\partial \epsilon^2)_{TP}}. \tag{9.6}$$

Let $(\partial V/\partial \epsilon) > 0$. Then $\Delta P > 0$ implies $\Delta \epsilon < 0$ by the preceding equation. Hence increasing the pressure causes the reaction to run backwards and thereby decrease the volume (relieving the pressure).

When external conditions are altered, the reaction always responds in such a way as to oppose the change (principle of Le Chatelier). If the response were opposite, a fluctuation in the constraints would be magnified by the reaction and lead to a runaway reaction, in contradiction with our assumption that the equilibrium is stable.

5.10   Other Applications

In this chapter the assumption of spatial homogeneity was dropped in order to discuss multiphase physical systems. Likewise one has to give up the assumption of isotropy to discuss solids; the relation between the stress and strain tensors then depends on the symmetry of the crystal, and the thermal equation of state becomes a tensor equation. Again in discussing interactions of matter with the electromagnetic field, there may be tensor relations between polarizations and fields. In other typical problems one may have additional constraints; for example, it may be necessary to express the electric neutrality of the system.

It is not possible to convey a proper feeling for the scope and power of thermodynamics without discussing applications at greater length and in more detail than has been attempted in these notes. At the end of this chapter a list of references to some of these applications has therefore been added [4–8].

Notes and References

1. J. W. Gibbs, *Collected Works*, Vol. 1 (Dover, New York, 1961).
2. The additive property of the entropy is also established by separating the mixture $A + B$ into parts A and B with the aid of the semi-permeable walls *aa* and *bb* shown in Figure 5.2. One moves the rigid cylinder *aacc* to the right. See, for example, M. Zemansky, *Heat and Thermodynamics*, Chapter 17 (McGraw-Hill, New York, 1959).

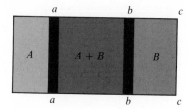

FIGURE 5.2

Separation of mixture $A + B$ into parts A and B. The walls *aa* and *bb* are permeable to A and B, respectively.

3. See, for example, Chapter 19 in Zemansky, reference [2], for an elementary, more detailed discussion.
4. P. S. Epstein, *Textbook of Thermodynamics* (Wiley, New York, 1937).
5. E. A. Guggenheim, *Thermodynamics* (North Holland, Amsterdam, 1949).
6. Herbert B. Callen, *Thermodynamics* (Wiley, New York, 1960).
7. P. M. Morse, *Thermal Physics* (Benjamin, New York, 1964).
8. L. D. Landau and E. M. Lifshitz, *Statistical Physics* (Addison-Wesley, Reading, Mass., 1958).

6

Statistical Foundations
of Thermodynamics

6.1   Introduction

According to thermodynamics the state of a closed, simple system in thermal equilibrium is entirely fixed by two variables, say P and V. On the other hand, from a molecular viewpoint, the number of degrees of freedom for systems of ordinary size is of the order of Avogadro's number. The thermodynamic formalism is thus an enormous abbreviation of the complete microscopic description. Our aim now is to see how the thermodynamic variables and the relations between them can be understood from the microscopic viewpoint.

The most basic fact that must be understood is the observed irreversible approach to thermodynamic equilibrium. Unfortunately, this fundamental question, which is sometimes called the Boltzmann problem [1], is still only partially answered. In addition we are excluding from the present volume systematic investigations of nonequilibrium situations, including such progress as has been made on the Boltzmann problem. Therefore we shall describe the provisional treatment which was discovered by Gibbs and permits us to postpone confrontation with the underlying questions. However, in section 6.13 and also in Chapter 10 there is some introductory discussion of the problem of the approach to thermodynamic equilibrium.

6.2   Phase Space

Since the thermodynamic method is applicable to an arbitrarily general
physical system, we shall study a correspondingly general system from the
microscopic viewpoint. Let the system be characterized by a set of generalized
coordinates and their equations of motion. It is most convenient to express
these equations in their canonical form

$$\frac{dp_k}{dt} = -\frac{\partial H}{\partial q_k} \qquad k = 1, \ldots, f$$

$$\frac{dq_k}{dt} = \frac{\partial H}{\partial p_k}$$

(2.1)

where H is the Hamiltonian function. For a system of interacting molecules
one has

$$H = \sum_i \mathbf{p}_i^2/2m + \sum_{i<j} V_{ij}. \tag{2.2}$$

For our general discussion, however, it will never be important to refer to
the exact form of the Hamiltonian.

It is of course not possible to solve the equations of motion exactly; it is
also impossible to determine the complete set $(q_1, \ldots p_1, \ldots)$ at any given
time t_0; so that even if the solutions were known, they could not be used.
Hence in order to give any useful description of a gas from first principles,
a statistical theory is unavoidable.

It is useful to introduce pq-space, phase space. There are two points of
view: (a) each molecule may be represented by a point in a six dimensional
phase space and the total gas is then represented by N points; (b) the complete
gas may be represented by a single point in a $6N$ dimensional phase space.

The first space, in which each molecule is separately represented, is called
μ-space. The second space, in which the whole gas is represented by a single
point, is termed Γ-space. We shall be concerned with Γ-space.

In order to give a statistical treatment we associate with the given physical
system an ensemble of points in Γ-space. Each of these points represents a
different possible state for the system. This set of points is called the Gibbs
ensemble (Figure 6.1).

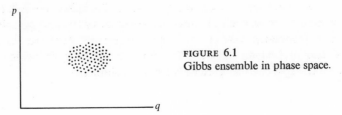

FIGURE 6.1
Gibbs ensemble in phase space.

Each representative point moves along a curve according to the Hamiltonian equations. It is assumed that such a curve is uniquely determined by a complete set $(\mathring{q}_1, \ldots \mathring{p}_1, \ldots)$ of initial conditions and therefore that these orbits can never intersect. Notice also that the points of a Gibbs ensemble contained within a closed hypersurface can never pass through that hypersurface if the surface itself moves according to (2.1).

6.3   Liouville Theorem

The motion of the ensemble is governed by the equations of motion (2.1) and also the equation of continuity,

$$\frac{\partial \rho}{\partial t} + \text{div} \, (\rho \, \mathbf{v}) = 0, \tag{3.1}$$

where ρ is the density of points in phase space and $\mathbf{v}$ is the velocity, with components:

$$\mathbf{v} = \left(\frac{dq_1}{dt} \cdots \frac{dq_f}{dt} \frac{dp_1}{dt} \cdots \frac{dp_f}{dt} \right). \tag{3.2}$$

We have from (3.1)

$$\rho \, \text{div} \, \mathbf{v} + \frac{D\rho}{Dt} = 0 \tag{3.3}$$

where

$$\frac{D\rho}{Dt} = \frac{\partial \rho}{\partial t} + (\mathbf{v} \cdot \text{grad}) \, \rho \tag{3.4}$$

is the hydrodynamical derivative, and

$$\text{div} \, \mathbf{v} = \sum_k \frac{\partial}{\partial q_k} \left(\frac{dq_k}{dt} \right) + \sum_k \frac{\partial}{\partial p_k} \left(\frac{dp_k}{dt} \right)$$

or

$$\text{div} \, \mathbf{v} = \sum_k \frac{\partial}{\partial q_k} \left(\frac{\partial H}{\partial p_k} \right) + \sum_k \frac{\partial}{\partial p_k} \left(-\frac{\partial H}{\partial q_k} \right) \tag{3.5}$$

$$= 0.$$

Hence

$$\frac{D\rho}{Dt} = 0. \tag{3.6a}$$

This is the Liouville theorem.

If we divide an ensemble into cells with the aid of a mesh which itself moves according to the Hamiltonian equations, then the number of points in each cell is conserved. Since the density is also conserved in the motion by Liouville's theorem, the volume of each cell and therefore of the entire ensemble is an integral of the motion. Hence

$$\frac{D\Omega}{Dt} = 0 \tag{3.6b}$$

where Ω is the volume of the ensemble.

Other forms of this theorem are

$$\frac{\partial \rho}{\partial t} = [H, \rho] \tag{3.6c}$$

where $[H, \rho]$ is the Poisson bracket, and

$$\frac{\partial(q_1, \ldots, p_n)}{\partial(q_{10}, \ldots, p_{n0})} = 1 \tag{3.6d}$$

which states that the Jacobian is volume preserving.

Equations (3.6c) and (3.6d) are the Eulerian and Lagrangian ways of describing the incompressibility of the flow.

6.4   Integrals of Motion

The volume of the ensemble may be called a collective integral of the motion. There are also other possible integrals associated with the motion of individual points, such as the integrals of energy, electric charge, angular momentum, etc. For example when the energy is conserved, each representative point is forced to move on a hypersurface, called the *ergodic* surface. If there are other integrals of the motion as well, e.g., the angular momentum, the paths are further restricted. In general the motion may be confined to a subregion of the ergodic surface because of the existence of special symmetries or integrals of the motion, or because of particular boundary conditions or constraints. Mathematical studies have shown that, except for special conditions such as we have just mentioned, there are no dynamical reasons for privileging particular regions of the ergodic surface. Unrestricted motions that come arbitrarily close to every point on the ergodic surface are called quasi-ergodic.

6.5   Steady State Ensembles and Thermodynamic Equilibrium

For a steady state one has

$$\frac{\partial \rho}{\partial t} = 0, \tag{5.1}$$

and by Liouville's theorem (3.6)

$$(\mathbf{v} \, \text{grad}) \, \rho = 0. \tag{5.2}$$

Hence grad $\rho = 0$ along a streamline, or ρ is constant along a streamline.

Assume that the motion is characterized by an energy integral and certain other integrals, say $\alpha_1, \alpha_2, \ldots$. Then the streamlines are confined to part of the ergodic surface as follows:

$$H\,(q_1, \ldots, q_f; p_1, \ldots, p_f) = E \tag{5.3}$$

$$\alpha_1\,(q_1, \ldots, q_f; p_1, \ldots, p_f) = a_1 \tag{5.4a}$$

$$\alpha_1\,(q_1, \ldots, q_f; p_1, \ldots, p_f) = a_2 \tag{5.4b}$$

$$\cdots$$

Since there appear to be no general dynamical reasons for assuming that the streamlines avoid particular sets of points in this subregion, we shall assume that ρ is constant throughout or

$$\rho = \rho\,(E, \alpha_1, \alpha_2, \ldots). \tag{5.5}$$

Such a density implies that all parts of phase space allowed by the external constraints and corresponding to given $E, \alpha_1, \alpha_2, \ldots$ are equally probable.

We shall postulate that such ensembles correspond to thermodynamic states. That is, it is assumed that no parts of phase space are dynamically preferred and that the weighting of particular parts of phase space can result only from special thermodynamic information, such as knowledge of the volume and total energy of the system. [2]

The most useful ensembles depend on the energy integral only and are of the following types:

Microcanonical: $\rho(E) = \delta(E - E_0)$ $\qquad$ (5.6)

Canonical: $\qquad$ $\rho(E) = e^{(\psi - E)/\theta}$ $\qquad$ (5.7)

Grandcanonical: $\rho(E) = e^{(\psi - E - \mu N)/\theta}$. $\qquad$ (5.8)

In the last case N is the number of particles.

6.6   Coarse-grained Density

An ensemble may be set up by measurements on a thermodynamically well-defined system. Since the thermodynamic description is incomplete, there are very many microscopic situations compatible with the given thermodynamic state. If all the microscopic coordinates could be exactly determined, one could construct the ensemble by making a series of measurements on the same thermodynamic state and recording one point in phase space after each measurement.

However, it is not possible to localize the system in phase space because of limitations on the precision of measurement. In classical theory these limitations are practical; in quantum theory they are fundamental.

The situation may be represented schematically by constructing a mesh in phase space (Figure 6.2), and introducing a coarse-grained density P_i as follows:

$$P_i \Delta\tau_i = \int \rho \, d\tau_i \tag{6.1}$$

$$\equiv N_i$$

where the integration is carried out over the ith cell, which has volume $\Delta\tau_i$. We suppose that the actual measurements yield a set of integers, $\{N_i\}$, where N_i is the number of times the system appears in the ith cell. For some purposes it is convenient to work with the continuous or fine-grained distribution (ρ), but of course only the coarse-grained or P distribution can be compared directly with experiment [3]. (One may imagine that the system is divided into small cells and that mass, energy, and momentum are measured in each cell. This information may then be transferred to Γ-space. The coarseness of this description is determined by the size of the cells.) One may normalize so that

$$\int \rho \, d\tau = 1 \tag{6.2a}$$

$$\sum_i P_i \Delta\tau_i = 1. \tag{6.2b}$$

Then expectation values may be defined as follows:

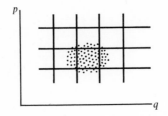

FIGURE 6.2
Coarse-graining in phase space.

$$\bar{\alpha} = \int \alpha \rho \, d\tau \tag{6.3a}$$

$$\bar{\alpha} = \sum_i \alpha_i P_i \Delta \tau_i. \tag{6.3b}$$

6.7   The Boltzmann *H*-function

The key function was introduced by Boltzmann. It is called the *H*-function and may be defined for both the fine- and coarse-grained ensemble [3]:

$$\bar{H} = \int \rho \ln \rho \, d\tau \tag{7.1a}$$

$$\bar{\bar{H}} = \sum_i P_i \ln P_i \Delta \tau_i. \tag{7.1b}$$

Suppose $\rho = $ constant. Then

$$\rho = 1/\Omega \tag{7.2}$$

where Ω is the volume of the ensemble. Then

$$\bar{H} = -\ln \Omega \tag{7.3a}$$

$$\bar{\bar{H}} = -\ln \Omega. \tag{7.3b}$$

If our knowledge of the thermodynamic system is relatively precise, the Gibbs ensemble will be highly localized and Ω will therefore be small. On the other hand, if little is known about the system, its Gibbs ensemble will be distributed over a large volume of phase space, and Ω will be large. Hence Ω^{-1} measures the information available and so also does

$$\ln \Omega^{-1} = -\ln \Omega = \bar{H} = \bar{\bar{H}} \tag{7.4}$$

for a uniform ensemble. The functions $\bar{H}$ and $\bar{\bar{H}}$ may be described as "information functions" and are equal only for uniform ensembles. When they are not equal, $\bar{H} > \bar{\bar{H}}$, as will now be shown.
    Consider

$$Q(\rho) = \rho \ln \frac{\rho}{P} - \rho + P. \tag{7.5}$$

Then $Q(P) = 0$. Also

$$\frac{dQ}{d\rho} = \ln \frac{\rho}{P}. \tag{7.6}$$

Hence

$$\left(\frac{dQ}{d\rho}\right)_{\rho=P} = 0 \tag{7.7a}$$

$$\left(\frac{dQ}{d\rho}\right) > 0 \quad \text{if } \rho > P \tag{7.7b}$$

$$\left(\frac{dQ}{d\rho}\right) < 0 \quad \text{if } \rho < P. \tag{7.7c}$$

Hence the function $Q(\rho)$ has a minimum at $\rho = P$ where it vanishes and therefore

$$Q(\rho) \geq Q(P) = 0 \tag{7.8}$$

and by integration over all of phase space we find

$$\int Q(\rho)\, d\tau \geq 0. \tag{7.9}$$

It follows from (7.5) and the preceding equation that

$$\int \rho \ln \frac{\rho}{P}\, d\tau \geq 0 \tag{7.10}$$

if

$$\int \rho\, d\tau = \int P\, d\tau, \tag{7.11}$$

as we shall assume. Another way to write (7.10) is

$$\int \rho \ln \rho\, d\tau \geq \int \rho \ln P\, d\tau. \tag{7.12}$$

The left hand side of (7.12) is just $\overline{H}$ and we shall now see that the right hand side is $\overline{\overline{H}}$. For

$$\int \rho \ln P\, d\tau = \sum_i \int_i \rho \ln P\, d\tau$$

$$= \sum_i \ln P_i \int_i \rho\, d\tau$$

$$= \sum_i P_i \ln P_i\, \Delta\tau_i$$

$$= \overline{\overline{H}}. \tag{7.13}$$

Hence

$$\overline{H} \geq \overline{\overline{H}}. \tag{7.14}$$

According to this result, if the same ensemble is described by functions $\overline{H}$

and $\bar{H}$, then $\bar{\bar{H}} \geq \bar{H}$. That is in agreement with the interpretation of $\bar{H}$ and $\bar{\bar{H}}$ as information functions. We may say that information is thrown away in passing from the fine-grained description ρ to the coarse-grained description P. This loss of information corresponds to a rounding off error.

6.8   Extensive Property of H

Let A and B be two noninteracting systems. Then

$$P_{A+B} = P_A P_B, \tag{8.1}$$

since states of the composite system $A + B$ are then obtained by just pairing states of the noninteracting parts in all possible ways. Then

$$\bar{\bar{H}}_{A+B} = \sum_{[A+B]} P_{A+B} \ln P_{A+B}$$

where the sum is over the phase space of the composite system. Then

$$\bar{\bar{H}}_{A+B} = \sum_{[A][B]} P_A P_B \ln P_A P_B$$

$$= \sum_{[A]} P_A \ln P_A + \sum_{[B]} P_B \ln P_B \tag{8.2}$$

where the sums are over the phase spaces of the separate systems. That is,

$$\bar{\bar{H}}_{A+B} = \bar{\bar{H}}_A + \bar{\bar{H}}_B. \tag{8.3}$$

Therefore $\bar{\bar{H}}$ and $\bar{H}$, like the entropy, are additive for noninteracting systems.

6.9   Dependence of $\bar{H}$ on Time

We have

$$\frac{D\rho}{Dt} = 0 \tag{9.1}$$

and

$$\frac{D\bar{H}}{Dt} = \frac{D}{Dt} \int \rho \ln \rho \, d\tau = \int \rho \, d\tau \left(\frac{D}{Dt} \ln \rho \right), \tag{9.2}$$

since

$$\frac{D}{Dt} (\rho \, d\tau) = 0. \tag{9.3}$$

Hence by (9.1) and (9.2)

$$\frac{D\bar{H}}{Dt} = 0 \tag{9.4}$$

or

$$\bar{H}(t_1) = \bar{H}(t_2). \tag{9.5}$$

Therefore the $\bar{H}$ measure of an ensemble is a constant of the motion.

6.10   Dependence of $\bar{\bar{H}}$ on Time (The $\bar{\bar{H}}$ Theorem)

The time dependence of $\bar{\bar{H}}$, on the other hand, cannot be inferred directly from the dynamical equations, since $\bar{\bar{H}}$ depends on P, not on ρ.

Assume that at some initial time (t_1) there is a measurement of the thermodynamical system which establishes $P_i(t_1)$. By hypothesis we cannot make a direct determination of $\rho(t_1)$. Instead we introduce the hypothesis

$$\rho(t_1) = P(t_1). \tag{10.1}$$

There are of course many ρ distributions compatible with the given (measured) P distribution. The simplest assumption is (10.1); any other single distribution would violate the principle of sufficient reason since it would favor certain regions of phase space over others in contradiction with our assumption that the only empirical information is entirely represented by P.

It follows from (10.1) that

$$\bar{\bar{H}}(t_1) = \bar{H}(t_1). \tag{10.2}$$

By (9.5),

$$\bar{H}(t_1) = \bar{H}(t_2). \tag{10.3}$$

By (7.14),

$$\bar{H}(t_1) \geq \bar{\bar{H}}(t_2). \tag{10.4}$$

The distinction between t_1 and t_2 is only this: at t_1 one assumes $\rho = P$; at t_2 this assumption is not made. One can say that at t_1 information is fed into the dynamical equations: $\bar{\bar{H}}(t_1) = \bar{H}(t_1)$; no information is gained or lost in using the dynamical equations: $\bar{H}(t_1) = \bar{H}(t_2)$; information may be lost in going from $\rho(t_2)$ to $P(t_2)$: $\bar{H}(t_2) \geq \bar{\bar{H}}(t_2)$; the inequality holds when there is a rounding off error.

6.11   Examples of Change in $\bar{\bar{H}}$ with Time

Consider an ideal gas confined to a definite volume with given total energy. Then

$$0 < x_i, y_i, z_i < a \tag{11.1}$$

$$E < \sum_i \mathbf{p}_i^2/2m < E + \Delta E. \tag{11.2}$$

The total available volume in phase space is

$$\Omega = \int \cdots \int d\mathbf{x}_1 \cdots d\mathbf{x}_N \, d\mathbf{p}_1 \cdots d\mathbf{p}_N \tag{11.3}$$

$$= \Omega_x \, \Omega_p \tag{11.4}$$

where

$$\Omega_x = \int \cdots \int d\mathbf{x}_1 \cdots d\mathbf{x}_N \tag{11.4a}$$

$$\Omega_p = \int \cdots \int d\mathbf{p}_1 \cdots d\mathbf{p}_N. \tag{11.4b}$$

Hence

$$\Omega_x = a^{3N} = V^N. \tag{11.5}$$

The volume of an n-dimensional sphere $\sim$ (radius)^n. Hence

$$\text{volume of sphere} \sim (2mE)^{3N/2}$$

$$\text{volume of shell} \sim E^{3N/2} \, \Delta E.$$

Therefore

$$\Omega_p \sim E^{3N/2} \, \Delta E. \tag{11.6}$$

We assume there is no other information. Hence the Gibbs ensemble uniformly fills the available phase space. Therefore $\bar{\Omega} = \bar{\bar{\Omega}}$ before the expansion, and

$$\bar{\bar{H}} = - \ln \bar{\bar{\Omega}} = - N \ln V - \frac{3N}{2} \ln E + \text{constant}. \tag{11.7}$$

In a free expansion we have

$$\Delta \bar{\bar{H}} = - N \ln \frac{V_2}{V_1}. \tag{11.8}$$

For natural processes, $V_2 > V_1$. Hence $\bar{\bar{H}}$ decreases. (Of course, $\bar{H}$ and $\bar{\Omega}$ are unchanged.)

In a thermal equalization we have initially

$$\sum_{1}^{N/2} \mathbf{p}_i^2 /2m = E_1$$

$$\sum_{\frac{N}{2}+1}^{N} \mathbf{p}_i^2 /2m = E_2 \tag{11.9}$$

and finally

$$\sum_{1}^{N} \mathbf{p}_i^2 /2m = E_1 + E_2. \tag{11.10}$$

In this example also we can show that $\bar{\Omega}_p$ (final) $> \bar{\Omega}_p$ (initial). In both of these examples

$$\Delta \bar{\bar{H}} = - \ln (\bar{\bar{\Omega}}_2/\bar{\bar{\Omega}}_1), \tag{11.11}$$

with $\bar{\bar{\Omega}}_2 > \bar{\bar{\Omega}}_1$ for natural processes. The first is an expansion in configuration space, and the second in momentum space.

6.12   Schematic Discussion of General Case

In general a thermodynamic state is characterized in phase space by a step function distribution P over a set of volumes which are defined by the macroscopic constraints. If we observe the time sequence of two thermodynamic states, we always find that an isolated system spontaneously evolves to the situation that offers fewer constraints. That is, in natural processes the volume $\bar{\bar{\Omega}}$ always expands and never contracts. We observe (Figure 6.3)

$$\text{(a)} \ A \longrightarrow B$$

and not

$$\text{(b)} \ B \longrightarrow A$$

In (a), the initial states (A) are dynamically connected with only a small fraction (B') of the final states (B). (The figure represents only the volume

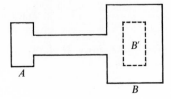

FIGURE 6.3
The initial region of phase space allowed by the thermodynamic constraints is A; the final region is B. The volume of B' equals the volume of A but has a very twisted shape. (B' is connected to A by the exact equations of motion.)

of the ensemble (B'); the actual shape of B' is some very twisted filament which wanders over all of B.) Then

$$\Omega(A) = \Omega(B') \ll \Omega(B).$$

In terms of the general proof, the transition $A \longrightarrow B'$ corresponds to the statement $\bar{H} = \bar{H}'$. The irreversibility is concentrated in the last step $(\bar{\bar{H}}' \leq \bar{H}')$ and has both a subjective and an objective aspect. On the subjective side one may say that information is lost by weakening the thermodynamic constraints. On the objective side there is the assumption, familiar to us from thermodynamics [4], that the system always takes advantage of the relaxed constraints. Implicit is the hypothesis that the structure of the system and its interaction with the heat bath are complex enough to provide mechanisms for occupying all phase space allowed by the constraints.

In the general proof, as in the specific examples of (6.11), one compares only the initial and final states. There is obviously no possibility of studying the evolution of the thermodynamic system from the initial to the final state. The conclusion is that a certain statistical function, $\bar{H}$, is permitted to change in only one direction. This is just the thermodynamic situation, and one argues again that the system tends as far in this preferred direction as the constraints allow.

To make practical use of this conclusion one then distributes those free physical quantities (e.g., mass, energy, momentum) in such a way as to make the appropriate thermodynamic or statistical function an extreme. Here one exactly parallels the thermodynamic analysis: for example the equation $dG \leq 0$ is always used in just this way.

6.13   Ergodic Theory and the Approach to Equilibrium

We have assumed that a thermodynamic state is represented by a Gibbs ensemble that covers the ergodic surface uniformly except for regions that it is forbidden to occupy because of macroscopic constraints. One adopts in particular a "uniform" ensemble because no part of phase space is dynamically preferred as far as we know. To adopt a nonuniform ensemble would be unjustified or arbitrary unless we had some special reason for favoring particular regions of phase space [2].

On the other hand the realistic situation is more as follows. Suppose that the isolated physical system, which is the subject of the thermodynamic analysis, has reached equilibrium so that its representative point will travel about the allowed region of phase space for an infinitely long time. Then the procedure of equal a priori probabilities would be incorrect if the representative point did not in fact spend equal times in equal areas of the ergodic surface. Therefore one would like to know that the following hypothesis is true:

$$\lim_{T \to \infty} \frac{t(A)}{T} = \frac{V(A)}{V} \tag{13.1}$$

where $t(A)$ is the time that the phase point spends in A and where $V(A)$ is the volume of A, while V is the total volume allowed by the constraints.

It has been possible to prove (13.1) for a certain kind of dynamical system, namely one which is designated "metrically transitive." A system of this type has the property that its energy surface cannot be divided into regions in such a way that orbits starting from points in one of the regions are permanently trapped in that region [5]. It has been objected that this result does not advance matters too much because of the difficulty in proving metrical transitivity—which appears very like the implicit assumption referred to in Section 6.12, i.e., that mechanisms are always available for occupying all of phase space allowed by the constraints. Nevertheless, as a matter of principle, one would like to follow the "ergodic program" in order to develop a unified approach to both equilibrium and nonequilibrium statistical mechanics [1, 8].

Here we present the common view that the quasi-ergodic hypothesis, i.e., the statement that almost every phase trajectory passes arbitrarily close to every point of the ergodic surface, is true but unproved for thermodynamically realistic systems. We adopt the same attitude toward (13.1)—the statement that the time average over the history of a single macroscopic system is equal to the result of averaging over an ensemble of systems similarly prepared.

We then have two pictures of a macroscopic system in thermodynamic equilibrium: either (a) an ensemble in phase space or (b) a single trajectory in phase space that covers the entire volume of (a). When the equilibrium is shifted, the system may depart from any point in the original ensemble, or any point on the original trajectory, and ultimately appear at any point in the final ensemble or the final trajectory. Since the initial and final constraints differ macroscopically, we are in no position to describe these external constraints microscopically; therefore the microscopic dynamical problem must be perturbed in a largely unknown way. Consequently the bundle of possible paths leading from the initial to the final microscopic state is also largely unknown.

One may say that $\overline{H}$ and $\overline{\overline{H}}$ are dynamical and thermodynamic functions respectively. In other words $\overline{\overline{H}}$ is chosen as a measure of the system because the thermodynamic description is incomplete. The irreversibility contained in $\overline{\overline{H}}$ arises formally when the equations of motion are partially replaced by stochastic assumptions. This mutilation of the exact reversible dynamics expresses the poor resolving power of thermodynamics—which does not distinguish small scale structure in either space or time and in addition misses many aspects of very long time behavior as well. Therefore the apparent or thermodynamic symmetry is less than the full dynamical symmetry and in particular no longer contains the symmetry element known as invariance under time reflections. The amount of irreversibility depends on the ratio $\Omega(B)/\Omega(B')$ (Figure 6.3), which is enormous for macroscopic systems. This is an example of "symmetry breaking" [6].

These matters would be much clearer if one could see the $\bar{\bar{H}}$-theorem emerge from an analysis of the approach to equilibrium, when the system is characterized by a realistic Hamiltonian and obeys the laws of quantum mechanics.

We shall return to the $\bar{\bar{H}}$-theorem in Chapter 10. Let us next investigate the consequences of the statement that $\bar{\bar{H}}$ tends toward a minimum.

6.14   Canonical Ensembles and Statistical Equilibrium

If we assume that $\bar{\bar{H}}$ is either constant or a decreasing function of time, then statistical equilibrium corresponds to a minimum in $\bar{\bar{H}}$. If the equilibrium is reached in such a way that the average ensemble energy is unchanged, we must satisfy the condition that $\bar{\bar{H}}$ is a minimum:

$$\delta\bar{\bar{H}} = 0, \tag{14.1}$$

subject to the constraint

$$\delta\bar{\bar{E}} = 0, \tag{14.2}$$

and also to the usual normalization

$$\sum_i P_i = 1. \tag{14.3}$$

Then

$$\delta\bar{\bar{H}} = \delta(\sum_i P_i \ln P_i)$$

$$= \sum_i \delta P_i \ln P_i \tag{14.4}$$

$$\delta\bar{\bar{E}} = \delta(\sum_i P_i E_i)$$

$$= \sum_i E_i \, \delta P_i. \tag{14.5}$$

By Lagrangian multipliers, we have

$$\sum_i (\ln P_i + aE_i + b) \, \delta P_i = 0$$

and

$$\ln P_i + aE_i + b = 0 \tag{14.6}$$

or

$$P_i = e^{(\psi - E_i)/\theta} \tag{14.7}$$

in the notation of Gibbs. Hence the canonical ensemble corresponds to statistical equilibrium. Therefore an arbitrary initial ensemble $P(E)$ will tend to evolve by successive interactions with its environment into a canonical one, provided that the ensemble energy is fixed. Obviously if there is no interaction with its environment, the system is isolated and the distribution $P(E)$ is unchangeable. We here assume slight interactions but subject to the condition that on the average the energy is not changed:

$$\delta \bar{E} = 0. \tag{14.2}$$

The restriction (14.2) is called the condition of essential isolation by Tolman [7]. This is the natural physical constraint for a system in equilibrium with a heat bath.

To connect the preceding formal proof with a realistic macroscopic system, imagine the system divided into small cells in which the usual physical quantities (mass, energy, momentum) are measured. This information may then be transferred to a coarse-grained ensemble which determines an initial $\bar{\bar{H}}$. If the system is not homogeneous, it will not be in equilibrium; and mass, energy, and momentum will redistribute themselves among the original cells. At any instant during this redistribution one could in principle measure the same physical variables and compute $\bar{\bar{H}}$. According to the formal discussion just given, $\bar{\bar{H}}$ trends downward to a minimum at which the ensemble is canonical.

6.15   Displacement of Equilibrium and the Second Law

We shall now investigate the displacement of equilibrium by changing the external constraints. This means for example changing the volume of the system or changing the strength of an applied magnetic field, and results in work on the system. The Hamiltonian now depends on the variables $q_1, \ldots, p_f$ and certain other parameters describing the constraints or interaction with the environment $a_1, \ldots, a_r$. If the $\{a_r\}$ are altered, we have

$$\Delta E_i = \sum_k \frac{\partial E_i}{\partial a_k} \Delta a_k \tag{15.1}$$

for the change in energy associated with a given cell in phase space.

The generalized forces may be defined by

$$A_k = -\frac{\partial E}{\partial a_k}.$$ (15.2)

Then

$$\Delta E_i = -\sum_k A_{ik}\,\Delta a_k.$$ (15.3)

We have for a canonical ensemble

$$\bar{\bar{H}} = \sum_i P_i \ln P_i$$

$$\Delta H = \sum_i \Delta P_i \ln P_i$$

$$= \sum_i \Delta P_i \left(\frac{\psi - E_i}{\theta}\right)$$

$$= -\frac{1}{\theta}\sum_i E_i\,\Delta P_i.$$ (15.4)

Therefore

$$-\theta\,\Delta H = \Delta\left(\sum_i E_i P_i\right) - \sum_i P_i\,\Delta E_i$$

$$= \Delta\bar{\bar{E}} - \overline{\overline{\Delta E}}$$ (15.5)

$$= \Delta\bar{\bar{E}} + \sum_k \bar{\bar{A}}_k\,\Delta a_k.$$ (15.6)

Since we want to associate statistical and thermodynamic equilibrium, the equation just given ought to be a statement of the second law:

$$T\,\Delta S = \Delta U + \Delta W.$$ (15.7)

Evidently also

$$\Delta U \longleftrightarrow \Delta\bar{\bar{E}}.$$ (15.8)

E_i has the meaning of the energy associated with given values $(q_1, \ldots, q_f\, p_1, \ldots, p_f)$. A change in E_i caused by a change in the external constraints is work done on or by the system. Hence

$$\Delta W \longleftrightarrow \overline{\overline{\sum_k A_k\,\Delta a_k}}.$$ (15.9)

Then we must have

$$k\,\Delta S \longleftrightarrow -\Delta H$$ (15.10)

and

$$kT \longleftrightarrow \theta. \tag{15.11}$$

Hence S corresponds to $-\bar{\bar{H}}$. The decreasing tendency of $\bar{\bar{H}}$ already found is consistent with the increasing property of the entropy. The additive property of $\bar{\bar{H}}$ for noninteracting systems corresponds to the extensive property of the entropy for such systems.

6.16   Other Thermodynamic Relations

Equation (15.6) is the analogue of the second law, which, according to the interpretation just given, relates two nearby canonical ensembles. The interpretation of ψ follows from the normalization condition:

$$\sum_i e^{(\psi - E_i)/\theta} = 1 \tag{16.1}$$

$$\psi = -\theta \ln Q \tag{16.2}$$

where

$$Q = \sum_i e^{-E_i/kT}. \tag{16.3}$$

Q is called the *partition function*. In addition

$$\bar{\bar{H}} = \sum_i P_i \ln P_i$$

$$= \frac{\psi - \bar{E}}{\theta}. \tag{16.4}$$

Hence

$$\psi \longleftrightarrow U - TS = A \tag{16.5}$$

and

$$A = -kT \ln Q. \tag{16.6}$$

The preceding formula is the most important practical link between statistical mechanics and thermodynamics. Q may in principle be calculated from the exact microscopic Hamiltonian with the aid of (16.3). Once A is known, the thermal and caloric equations of state are given as follows:

$$P = -\left(\frac{\partial A}{\partial V}\right)_T \tag{16.7}$$

$$S = -\left(\frac{\partial A}{\partial T}\right)_V. \tag{16.8}$$

All thermodynamic properties are then in principle deducible from the exact microscopic Hamiltonian.

Instead of (16.8) the following is often useful:

$$U = A + TS$$

$$= A - T\left(\frac{\partial A}{\partial T}\right)$$

or

$$U = -T^2 \frac{\partial}{\partial T}\left(\frac{A}{T}\right). \tag{16.9}$$

The compatibility condition between the thermal and caloric equation is satisfied since

$$\left(\frac{\partial P}{\partial T}\right)_V = \left(\frac{\partial S}{\partial V}\right)_T = -\frac{\partial^2 A}{\partial V \, \partial T}. \tag{16.10}$$

6.17   Molecular Distributions

In the present analysis the entire physical system is represented by a single point in phase space (Γ-space). However, it is also possible to infer information about the distributions of the individual molecules from the same Γ-space distribution.

MAXWELL-BOLTZMANN DISTRIBUTION

If the ensemble is canonical, the probability that the gas as a whole has the energy E is

$$P = e^{(\psi - E)/\theta}. \tag{17.1}$$

The distribution of any particular molecule or group of molecules over phase space may be calculated by integrating over the coordinates of all the other molecules. Thus the probability that a particular molecule (1) lies in the cell $d\mathbf{x}_1 \, d\mathbf{p}_1$ is

$$(d\mathbf{x}_1 \, d\mathbf{p}_1) \int \cdots \int d\mathbf{x}_2 \cdots d\mathbf{x}_N \, d\mathbf{p}_2 \cdots d\mathbf{p}_N \, e^{(\psi - E)/\theta}. \tag{17.2}$$

The Hamiltonian may be assumed to have the form

$$H = \sum_i \epsilon_i(\mathbf{p}_i) + \sum_{i<j} V_{ij}(|\mathbf{x}_i - \mathbf{x}_j|). \tag{17.3}$$

Then the probability that the first molecule lies in the cell $d\mathbf{p}_1$ is

$$d\mathbf{p}_1 \left[\exp\left(\psi - \epsilon_i\right)/\theta \right] \int \cdots \int d\mathbf{p}_2 \cdots d\mathbf{p}_N \, d\mathbf{x}_1 \cdots d\mathbf{x}_N$$

$$\exp\left(-\left[\sum_{i=2}^{N}\epsilon_i + \sum_{i<j} V_{ij}\right]/kT\right) \quad (17.4)$$

or

$$A\, e^{-p^2/2mkT}\, d\mathbf{p} \quad (17.5)$$

with our previous interpretation of θ. Thus if the Gibbs ensemble is canonical, the individual molecules are distributed in momentum space according to the Maxwell-Boltzmann distribution.

CORRELATION COEFFICIENTS

The probability that molecules 1 and 2 are at positions x_1 and x_2, independent of where the other particles are and of how fast they are moving, is

$$W(\mathbf{x}_1, \mathbf{x}_2) = \left[\int \cdots \int d\mathbf{p}_1 \cdots d\mathbf{p}_N \int \cdots \int d\mathbf{x}_3 \cdots d\mathbf{x}_N \right.$$

$$\left. P(\mathbf{x}_1 \cdots \mathbf{x}_N\, \mathbf{p}_1 \cdots \mathbf{p}_N)\right]\, d\mathbf{x}_1\, d\mathbf{x}_2. \quad (17.6)$$

Other correlations may be similarly expressed.

EQUIPARTITION THEOREM

The normalization condition may be written

$$\int \cdots \int A e^{-E/\theta}\, dq_1 \cdots dq_f\, dp_1 \cdots dp_f = 1 \quad (17.7)$$

for a system with f degrees of freedom. We have

$$\int e^{-E/\theta}\, dq_1 = \left[q_1\, e^{-E/\theta}\right] + \frac{1}{\theta}\int q_1 \frac{\partial E}{\partial q_1} e^{-E/\theta}\, dq_1. \quad (17.8)$$

For any localized system, the first term vanishes and

$$\int e^{-E/\theta}\, dq_1 = \frac{1}{\theta}\int q_1 \frac{\partial E}{\partial q_1} e^{-E/\theta}\, dq_1.$$

But the expectation value of any observable α is

$$\bar{\alpha} = A \int \cdots \int \alpha e^{-E/kT}\, dq_1 \cdots dp_f.$$

Hence

$$\frac{1}{\theta}\overline{\frac{\partial E}{\partial q_1} q_1} = 1$$

or

$$\overline{\frac{\partial E}{\partial q_1}} q_1 = kT. \tag{17.9}$$

Similarly,

$$\overline{\frac{\partial E}{\partial p_1}} p_1 = kT \tag{17.10}$$

if the system is localized in momentum space. If the kinetic energy has the familiar quadratic dependence on p_i, we have

$$2\overline{E} = \sum_{i=1}^{f} \overline{p_i \frac{\partial E}{\partial p_i}} = fkT$$

$$\overline{E} = \left(\frac{1}{2} kT\right) f \tag{17.11}$$

where f is the number of degrees of freedom. In general the equipartition theorem is expressed by equations (17.9) and (17.10). The energy $\frac{1}{2}kT$ is associated with each degree of freedom in the way indicated by these equations.

6.18   Grand Canonical Distribution

Molecular distributions (μ-space distributions) contain less information than the canonical distribution in Γ-space. It is alternatively possible to ascend to the grand canonical distribution which contains more information than the canonical.

One may modify the argument leading to the canonical distribution by treating the number of particles in the same way as the total energy. Then there are two constraints:

$$\delta \overline{E} = \delta \left(\sum_i P_i E_i\right) = 0 \tag{18.1}$$

$$\delta \overline{\overline{N}} = \delta \left(\sum_i P_i N_i\right) = 0 \tag{18.2}$$

and, correspondingly, an additional Lagrangian multiplier. The result is

$$P_i = e^{(\psi - E_i - \mu N_i)/\theta}. \tag{18.3}$$

The grand canonical distribution will be discussed in Chapter 8.

6.19   Adaptation of Argument to Quantum Theory

Since atoms obey the laws of quantum mechanics, Gibbs was unsuccessful in applying his formalism to real physical systems. However, because of the very deep correspondence between classical and quantum mechanics, the classical arguments about the foundations of thermodynamics are very little changed by quantum theory.

On the other hand, as a matter of principle, classical statistics should be obtained from quantum statistics by the correspondence principle, and not vice versa. Here we shall first go over from the classical to the quantum formalism by changing a single equation, namely the formula for the partition function. We may then regard our new quantum formula for the partition function as a postulate and derive from it the classical formalism.

In Section 6.21 it will be shown in a more satisfactory way how macroscopic theory may be based entirely on quantum mechanical concepts. There the quantum partition function will be obtained in terms of the so-called density matrix. However, as long as thermodynamic equilibrium is not satisfactorily obtained by solving the time dependent equations, one is in fact really postulating the partition function, both classically and quantum mechanically.

Accordingly we now postulate the following exact connection between microscopic and macroscopic theory. Let the time independent Schrodinger equation of the complete physical system be

$$H\Psi = E\Psi \tag{19.1}$$

(Here H is the exact Hamiltonian, e.g., for a system of structureless particles,

$$H = \sum_i \mathbf{p}_i^2/2m + \sum_{i<j} V_{ij} \tag{19.2}$$

where $\mathbf{p}_i = \dfrac{\hbar}{i} \boldsymbol{\nabla}_i$.)

Let the energy spectrum be $E(a_1, a_2, \ldots)$ where $a_1, a_2, \ldots$ is the complete set of quantum numbers. Next define the partition function to be

$$Q = \sum_{a_1, a_2, \ldots} e^{-E(a_1, a_2, \ldots)/kT} \tag{19.3}$$

where the sum extends over the complete spectrum of Schrodinger's equation (19.1). We finally postulate that the connection with thermodynamics is not altered. Then

$$A = -kT \ln Q \tag{19.4}$$

$$P = -\left(\frac{\partial A}{\partial V}\right)_T \tag{19.5}$$

$$S = -\left(\frac{\partial A}{\partial T}\right)_V. \tag{19.6}$$

There is thus a simple and unambiguous connection in principle between the Schrodinger equation and the thermal and caloric equations of state. In fact from knowledge of atomic structure alone it is possible in principle to calculate all thermodynamic properties. For example, knowing the structure of hydrogen in terms of electrons and protons and taking into account only the Coulomb field between them, it is possible in principle by equations (19.1)–(19.6) to determine the equations of state of all the separate phases of hydrogen as well as the phase transitions.

The sum (19.3) may also be abbreviated

$$Q = \sum_{states\,(\alpha)} e^{-E_\alpha/kT} \tag{19.7}$$

where this notation means a sum over all states. When two or more states belong to the same energy level, that energy level is called degenerate. Let g_n be the degeneracy of the level E_n. Then

$$Q = \sum_{\substack{n \\ (levels)}} g_n\, e^{-E_n/kT} \tag{19.8}$$

where the sum is now over levels.

An alternative notation is

$$Q(\lambda) = \int_0^\infty e^{-\lambda E}\, \rho(E)\, dE \tag{19.9}$$

where $\lambda = 1/kT$ and $\rho(E)\, dE$ is the number of states (the number of eigensolutions of Schrodinger's equation) in the corresponding energy interval.

6.20   Third Law

At the lowest temperatures the leading term in the partition function is

$$Q \cong g_0\, e^{-E_0/kT} \tag{20.1}$$

where E_0 is the ground state of the physical system. Then

$$A \cong -kT \ln g_0 + E_0 \tag{20.2}$$
$$S \cong k \ln g_0. \tag{20.3}$$

It follows that the limit of the entropy near $T = 0$ is given by

$$\lim_{T\to 0} S = k \ln g_0. \tag{20.4}$$

The question of whether $S(T = 0)$ depends on other physical parameters depends in turn on whether the degeneracy g_0 is removed by these other parameters. If there is a degeneracy g_0 that could be removed by a magnetic field, for example, then the third law in the Nernst form would not be true.

Therefore the Nernst form, if correct, implies either that the degeneracy of the ground state is not directly observable since it cannot be removed, or that the ground state is always nondegenerate. It is possible to conjecture that $g_0 = 1$ always and therefore to formulate the third law as simply

$$S_0 = 0. \tag{20.5}$$

This is the Planck formulation.

6.21   Density Matrix and Quantum Partition Function

The connection between atomic structure and thermodynamics is precisely given in (19.3), the partition function. Our main interest is in applications of this relation rather than in derivations of it, since such derivations lie in the province of nonequilibrium statistical mechanics. Nevertheless we have given a classical introduction to this relation and would now like to sketch the corresponding quantum argument, which is formally, but not conceptually, very similar. That is, we shall obtain the quantum partition function from the density matrix just as the classical partition function was obtained from the density in phase space.

In quantum theory a physical state is no longer represented by a point in phase space but rather by a point in Hilbert space, and a Gibbs ensemble becomes a distribution in Hilbert space. To describe this distribution one introduces a density matrix and an $\overline{H}$ function. Following the classical pattern, one also introduces a coarse-grained density and a corresponding coarse-grained $\overline{\overline{H}}$; there is an $\overline{H}$ theorem, and canonical ensembles again correspond to thermodynamic equilibrium.

QUAMTUM DESCRIPTION OF THERMODYNAMIC STATE

Let the physical system be represented by the state function, ψ, and let ψ be expanded in some complete set $\psi(a')$ of basis functions. Then

$$\psi = \sum_{a'} c(a')\,\psi(a'). \tag{21.1a}$$

If the $\psi(a')$ are the eigenfunctions of some observable, then $|c(a')|^2$ is the probability of observing the particular eigenvalue a' when the system is prepared in the state ψ. In Dirac notation

$$|\rangle = \sum_{a'} |a'\rangle \langle a'| \rangle \tag{21.1b}$$

where

$$\psi = |\rangle, \psi(a') = |a'\rangle, \quad \text{and} \quad c(a') = \langle a'|\rangle.$$

In classical theory a precisely specified system is defined by its coordinates and momenta, or by a single point in phase space. In quantum theory, on the other hand, a completely defined system is entirely characterized by the state function, ψ, or by the point in function space with coordinates $c(a')$.

A single thermodynamic state corresponds to a large number of micro-states. In classical theory these are represented by a Gibbs ensemble in phase space and in quantum theory by an ensemble in function space (or state space). Let the weight (probability) of the volume element $d\tau$ be $\rho\,d\tau$ in classical theory, and let the weight of the discrete state (α) be w_α in quantum theory. Then the following formulas hold in quantum and classical theory, and are to be used for ensembles representing thermodynamic states:

$$\langle A \rangle = \int A\,\rho\,d\tau \qquad \text{classical} \qquad (21.2)$$

$$\langle A \rangle = \sum_\alpha w_\alpha \langle \alpha | A | \alpha \rangle \qquad \text{quantal} \qquad (21.3)$$

when $\langle A \rangle$ is the expectation value of the observable A. These formulas imply the normalizations

$$\int \rho\,d\tau = 1 \qquad \text{classical} \qquad (21.4)$$

$$\sum_\alpha w_\alpha = 1 \qquad \text{quantal.} \qquad (21.5)$$

A pure state is defined by the condition $w_\alpha = \delta_{\alpha 1}$. The corresponding ensemble degenerates to a single point in state space, namely $|1\rangle$.

DENSITY MATRIX

Let us consider some other set of basis functions, say $|s\rangle$. Then

$$|\alpha\rangle = \sum_s |s\rangle \langle s | \alpha \rangle$$

and

$$\langle \alpha | = \sum_s \langle \alpha | s \rangle \langle s |.$$

Therefore (21.3) becomes

$$\langle A \rangle = \sum_{\alpha, s, s'} w_\alpha \langle \alpha | s \rangle \langle s | A | s' \rangle \langle s' | \alpha \rangle. \qquad (21.6)$$

Define the density matrix, ρ, by the equation

$$\langle s' | \rho | s \rangle = \sum_\alpha \langle s' | \alpha \rangle w_\alpha \langle \alpha | s \rangle \qquad (21.7)$$

summed over the ensemble in state space. Then

$$\langle A \rangle = \sum_{s,s'} \langle s'|\rho|s \rangle \langle s|A|s' \rangle$$

or

$$\langle A \rangle = \mathrm{Tr}\,\rho A$$
$$= \mathrm{Tr}\,A\rho \tag{21.8}$$

where Tr means trace. In addition

$$\sum_{s} \langle s|\rho|s \rangle = \sum_{s,\alpha} \langle s|\alpha \rangle\, w_\alpha \langle \alpha|s \rangle$$
$$= \sum_{\alpha} w_\alpha = 1.$$

Therefore

$$\mathrm{Tr}\,\rho = 1. \tag{21.9}$$

Note also that if ρ describes a pure state,

$$\rho^2 = \rho. \tag{21.10}$$

LIOUVILLE THEOREM

We may calculate the time dependence of ρ as follows. We have from (21.7)

$$\rho = \sum_{\alpha} |\alpha \rangle\, w_\alpha \langle \alpha|,$$

and by the Schrodinger equation,

$$i\hbar \frac{\partial}{\partial t} |\alpha \rangle = H|\alpha \rangle.$$

Then

$$i\hbar \frac{\partial \rho}{\partial t} = \sum_{\alpha} (H|\alpha \rangle\, w_\alpha \langle \alpha| - |\alpha \rangle\, w_\alpha \langle \alpha|H)$$

and therefore

$$i\hbar \frac{\partial \rho}{\partial t} = H\rho - \rho H$$

or

$$i\hbar \frac{\partial \rho}{\partial t} = (H, \rho). \tag{21.11}$$

The last equation is related to the Poisson bracket form of the classical equation (3.6c) of this chapter by the usual substitution:

$$[\, , \,] \rightarrow (\, , \,)/i\hbar.$$

The condition for statistical equilibrium is then

$$(H, \rho) = 0. \tag{21.11a}$$

If (21.11a) is satisfied, then ρ may be taken diagonal in the energy representation.

CONFIGURATION SPACE

In the x-representation the density matrix becomes

$$\langle x|\rho|x'\rangle = \sum \langle x|\alpha\rangle \, w_\alpha \, \langle \alpha|x'\rangle. \tag{21.12a}$$

In the preceding equation, x is an abbreviation for the complete set $(x_1, \ldots, x_n)$ if there are N-particles. In this representation the diagonal elements are simply related to the probability density:

$$\langle x_1, \ldots, x_N|\rho|x_1, \ldots, x_N\rangle = \sum_\alpha |\langle x_1, \ldots, x_N|\alpha\rangle|^2 \, w_\alpha$$

$$= \sum_\alpha w_\alpha \, |\psi_\alpha(x_1, \ldots, x_N)|^2. \tag{21.12b}$$

QUANTUM MECHANICAL H-FUNCTION

We define

$$\bar{H} = \sum_n \rho_{nn} \ln \rho_{nn}. \tag{21.13}$$

In Appendix D it is shown that

$$\bar{H}(0) \geq \bar{H}(t), \tag{21.14}$$

provided that $\rho(0)$ is diagonal. Note that this equation differs from the classical results by the inequality sign. One may say that the classical $\bar{H}$-function conserves information while in quantum mechanics there is a corresponding loss of information.

COARSE-GRAINING

Because the thermodynamic description is incomplete, coarse-graining is again required. One defines just as in the classical case:

$$P_N = \frac{1}{g_N} \sum_{n=1}^{g_N} \rho(n, n), \tag{21.15}$$

and a coarse-grained H-function,

$$\overline{\overline{H}} = \sum_N P_N \ln P_N \tag{21.16}$$

where g_N is the number of microscopic states that are not distinguishable at the given level of observational precision. One again proves

$$\overline{\overline{H}} \le \overline{H}. \tag{21.17}$$

The discussion continues to parallel the classical argument with the same conclusion, namely:

$$\overline{\overline{H}}(t_2) \le \overline{\overline{H}}(t_1) \qquad t_2 > t_1. \tag{21.18}$$

This result depends on the assumption [needed to prove (21.14)] that $\rho(0)$ is diagonal. This assumption in turn is equivalent to a random phase postulate and may be regarded as supplementing at the quantum level the hypothesis of equal a priori probabilities (see Appendix D).

CANONICAL ENSEMBLE

Finally one is led to associate canonical ensembles with thermodynamic equilibrium and therefore to recover the classical formula for the partition function except that the spectrum is now determined quantum mechanically (for example, by the Schrodinger equation). Thus one derives a coarse-grained version of (19.7).

The corresponding coarse-grained density is

$$P_n = e^{(\psi - E_n)/\theta}. \tag{21.19}$$

QUANTAL AND CLASSICAL LIMITS

The expressions (21.7) and (21.12) apply to an arbitrary macroscopic system and therefore embrace both classical and quantal behavior. The classical limit is characterized by a high degree of incoherence and is expressed by a wide spread, which grows larger as the temperature is increased, over the weights w_α. The quantum limit, on the other hand, is characterized by a single quantum mechanical state and therefore complete coherence. Formally the quantum limit is described by the condition:

$$\rho^2 = \rho.$$

The third law expressed in terms of the density matrix then reads

$$\lim_{T \to 0} (\rho^2 - \rho) = 0.$$

Notes and References

1. G. E. Uhlenbeck and G. W. Ford, *Statistical Mechanics* (American Mathematical Society, Providence, 1963), p. 125.
2. This point of view is Tolman's hypothesis of equal a priori probabilities. See R. C. Tolman, *The Principles of Statistical Mechanics* (Clarendon Press, Oxford, 1938).
3. The necessity of coarse-graining was pointed out by the Ehrenfests. See reference [2] and P. and T. Ehrenfest, *Conceptual Foundations of the Statistical Approach in Mechanics*, translated by M. J. Moravcsik (Cornell Univ. Press, Ithaca, 1949).
4. See remarks in Section 3.5, for example.
5. G. D. Birkhoff, *Proc. Nat. Acad.* Sci. U. S. 17: 656 (1931). A. I. Khinchin, *Mathematical Foundations of Statistical Mechanics*, translated by G. Gamow (Dover, New York, 1949).
6. Another kind of symmetry breaking is described in Chapter 10. In both cases the complete symmetry of the Hamiltonian is broken in the sense that under the given physical conditions one observes only a subset of all properties of the system. In Chapter 10, one examines a system in the solid phase, for example, and observes only crystallographic symmetry, instead of the full symmetry of the Hamiltonian, which is invariant under the complete rotation-translation group.
7. See reference [2], p. 498.
8. Important progress has recently been made by Ja Sinai who has proved that the classical motion of N particles ($N \geq 3$) with short range repulsive forces (e.g., hard spheres) is metrically transitive. See *Statistical Mechanics*, edited by T. A. Bak (Benjamin, New York, 1967).

7

Applications to Some Simple Systems

7.1 Ideal Systems

We shall call a system ideal if its energy has the following simple structure as a function of the quantum numbers (a_i):

$$E = \sum_i E_i(a_i) \qquad (1.1)$$

where the sum is over (noninteracting) parts (i). Examples of ideal systems are ideal gases, Einstein and Debye crystals, and paramagnets. If the energy is additive as shown, then the partition function is multiplicative:

$$Q = \sum_{a_1, a_2, \ldots} e^{-[E(a_1)+E(a_2)+\cdots]/kT}$$

$$= \sum_{a_1} e^{-E(a_1)/kT} \sum_{a_2} e^{-E(a_2)/kT} \cdots \qquad (1.2)$$

or

$$Q = \Pi \, Q_i \qquad (1.3)$$

where

$$Q_i = \sum_{a_i} e^{-E_i(a_i)/kT}. \qquad (1.3a)$$

Then

$$A = \sum_i A_i \tag{1.4}$$

where

$$A_i = - kT \ln Q_i \tag{1.4a}$$

Clearly pressures and entropies of noninteracting parts are also additive, by equations (19.5) and (19.6) of Chapter 6, and equation (1.4) above.

7.2   System of Oscillators

EINSTEIN CRYSTAL

An Einstein crystal is composed of noninteracting oscillators, all of which have the same frequency; the individual oscillators are in this case the nuclei oscillating about their equilibrium positions.

Since the noninteracting parts are in this example identical, we have

$$A = NA_1 \tag{2.1}$$

$$A_1 = - kT \ln Q_1 \tag{2.1a}$$

where Q_1 is the partition function of a single oscillator:

$$Q_1 = \sum_i e^{-E_i/kT}$$

$$= Q_x^3 \tag{2.2}$$

where

$$Q_x = Q_y = Q_z = \sum_n e^{-\left(n+\frac{1}{2}\right)\frac{\hbar\omega}{kT}}$$

$$= e^{-\frac{1}{2}\frac{\hbar\omega}{kT}} \sum_n x^n,$$

and

$$x = e^{-\hbar\omega/kT}.$$

Therefore

$$Q_x = \frac{x^{1/2}}{1 - x}$$

$$2Q_x = \frac{1}{\sinh \dfrac{\hbar\omega}{2kT}} \tag{2.3}$$

Hence

$$A = 3NkT \ln 2 \left(\sinh \frac{\hbar\omega}{2kT} \right). \tag{2.4}$$

POLYATOMIC MOLECULES

If the different independent oscillators do not all have the same frequency, the preceding result is only slightly changed, namely:

$$A = kT \sum_i \ln 2 \sinh \frac{\hbar\omega_i}{2kT} \tag{2.5}$$

where the sum is over all oscillators. The preceding formula holds, for example, for a polyatomic molecule; in this application the ω_i have to be understood as frequencies of the normal vibrational modes.

SYSTEM OSCILLATING NEAR EQUILIBRIUM (LINEAR SYSTEM)

Let us consider any dynamical system oscillating about an equilibrium. It is then possible to expand the potential energy about the equilibrium configuration. Provided that the system is adequately described by linear equations of motion, the potential energy is given accurately enough by quadratic terms in the Taylor's expansion:

$$V = \sum_{ij} a_{ij} q_i q_j \tag{2.6}$$

where q_i are the generalized coordinates. In problems of this type it is usual to transform to coordinates that reduce V to a sum of squares without at the same time altering the form of kinetic energy:

$$T = \sum_i \frac{\mu_i}{2} \dot{q}_i^2. \tag{2.7}$$

This may in general be done by a linear orthogonal transformation to new coordinates:

$$Q_i = \sum_j c_{ij} q_j. \tag{2.8}$$

In terms of these new coordinates, Q_i, which are called normal coordinates, the total Hamiltonian may be written as a sum of the following type:

$$H = \sum_i H_i \tag{2.9}$$

where

$$H_i = \frac{1}{2M_i} (P_i^2 + M_i^2 \, \omega_i^2 \, Q_i^2). \tag{2.9a}$$

The contribution H_i may be interpreted as the Hamiltonian of an oscillator with the effective mass M_i and frequency ω_i. The eigenvalues of the total energy are

$$E = \sum_i \left(n_i + \frac{1}{2} \right) \hbar \omega_i. \tag{2.10}$$

The work of the preceding paragraph is therefore applicable and the free energy is

$$A = kT \sum_i \ln 2 \sinh \frac{\hbar \omega_i}{2kT} \tag{2.11}$$

where the sum is extended over all normal modes. Equation (2.11) is valid for any linear system oscillating about an equilibrium.

7.3   Debye Crystal (Caloric Equation)

If the system has a nearly continuous spectrum, A may be approximated by an integral:

$$A = kT \int_0^\infty \ln \left(2 \sinh \frac{\hbar \omega}{2kT} \right) \rho(\omega) \, d\omega \tag{3.1}$$

where $\rho(\omega) \, d\omega$ is the number of modes in the interval ω, $\omega + d\omega$. Here $\rho(\omega)$ is subject to the normalization condition

$$\int_0^\infty \rho(\omega) \, d\omega = f \tag{3.2}$$

where f is the total number of degrees of freedom.

In the case of a crystal with N nuclei,

$$f = 3N. \tag{3.2a}$$

The spectrum of a crystal may be divided into two parts as shown in Figure 7.1. For low frequencies,

$$\rho(\omega) = a\omega^2 \tag{3.3}$$

where a depends on the crystal but where the quadratic dependence is the same for all crystals. The high frequency part of the spectrum is structure dependent since in this region the wave length of sound approaches the order of the lattice spacing and can therefore distinguish crystal structure.

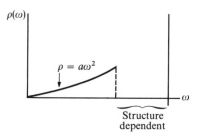

FIGURE 7.1
Frequency spectrum of a crystal.

The Debye model of a crystal approximates $\rho(\omega)$ by $a\omega^2$ over the whole spectrum. It is therefore a one parameter (a) model based on the part of the spectrum that has the same form for all crystals. In order to satisfy the normalization condition such a quadratic spectrum has to be cut off:

$$\int_0^{\bar{\omega}} a\omega^2 \, d\omega = 3N$$

or

$$a\bar{\omega}^3 = 9N. \tag{3.4}$$

The cutoff $\bar{\omega}$ may be regarded as a crude way to take into account the lattice structure of the crystal.

One may relate a to the transverse and longitudinal velocities of sound (c_t, c_l) by regarding the crystal as a continuum enclosed in a certain volume (V). Then one finds

$$a = 4\pi\left(\frac{1}{c_l^3} + \frac{2}{c_t^3}\right)\frac{V}{(2\pi)^3}. \tag{3.5}$$

The free energy now becomes, by (3.1) and (3.3), and independent of (3.5),

$$A = akT \int_0^{\bar{\omega}} \ln\left(2 \sinh\frac{\hbar\omega}{2kT}\right) \omega^2 \, d\omega. \tag{3.6}$$

Let

$$x = \hbar\omega/kT. \tag{3.7}$$

Define also the Debye temperature (Θ) by

$$\hbar\bar{\omega} = k\,\Theta. \tag{3.8}$$

Then

$$A = 9R\,\Theta\left(\frac{T}{\Theta}\right)^4 G\left(\frac{\Theta}{T}\right) \tag{3.9}$$

where

$$G(\bar{x}) = \int_0^{\bar{x}} F(x) \, x^2 \, dx \qquad (3.9a)$$

$$F(x) = \ln\left(2 \sinh \frac{x}{2}\right). \qquad (3.9b)$$

The specific heat calculated from this form of A is called the Debye specific heat; near absolute zero it varies as T^3. We have

$$F(x) = \ln(e^{x/2} - e^{-x/2})$$

$$= \ln e^{x/2}(1 - e^{-x})$$

$$= \frac{x}{2} + \ln(1 - e^{-x})$$

$$\frac{A}{9R\,\Theta} = \frac{1}{8} + \left(\frac{T}{\Theta}\right)^4 \int_0^{\Theta/T} \ln(1 - e^{-x}) \, x^2 \, dx. \qquad (3.10)$$

Near absolute zero, one may approximate by putting $\Theta/T = \infty$ in the upper limit of the integral. Then

$$\frac{A}{9R\,\Theta} \simeq \frac{1}{8} + B\left(\frac{T}{\Theta}\right)^4 \qquad (3.10a)$$

$$B = \int_0^\infty \ln(1 - e^{-x}) \, x^2 \, dx = -\frac{\pi^4}{45}. \qquad (3.10b)$$

For C_V one finds

$$C_V = -T\left(\frac{\partial^2 A}{\partial T^2}\right)_V$$

$$= -108B\left(\frac{T}{\Theta}\right)^3$$

$$C_V = \frac{12\pi^4}{5}\left(\frac{T}{\Theta}\right)^3. \qquad (3.11)$$

At low temperatures only the long wave lengths are excited. This part of the frequency spectrum is structure independent and therefore at low temperatures all crystals behave alike, i.e., they all exhibit the T^3 temperature dependence of the specific heat.

By investigating the specific heat in the high temperature limit (where the temperature is as high as possible without violating the model), one finds again that the result is the same for all kinds of crystals; in this case one recovers the classical value

$$C_V = 3R. \qquad (3.12)$$

Thus the low and "high" temperature limits take the same form for all crystals. (See Figure 3.1 of Chapter 3.)

7.4   Debye Crystal (Thermal Equation)

The pressure of the Debye waves may be calculated from the usual equation:

$$P_D = - \left(\frac{\partial A_D}{\partial V}\right)_T.$$

(4.1)

We shall abbreviate A_D as follows:

$$A_D = Tf\left(\frac{\Theta}{T}\right).$$

(4.1a)

Then

$$P_D = - \frac{d\Theta}{dV}f'$$

where f' is the derivative of f with respect to its argument. For the internal energy, we find

$$U_D = A_D - T\left(\frac{\partial A_D}{\partial T}\right)_V$$

or

$$= \Theta f'$$

Hence

$$P_D = - \frac{d\Theta}{dV}\frac{U_D}{\Theta}$$

$$= \gamma \frac{U_D}{V}$$

(4.2)

where

$$\gamma = - \frac{d \ln \Theta}{d \ln V}.$$

(4.2a)

One finds empirically that $\gamma \cong 2$ (approximately temperature independent). One may compare this value of γ with that appropriate to nonrelativistic and relativistic gases. If we write there also that the pressure is proportional to the energy density as follows:

$$P = \gamma \frac{U}{V},$$

(4.3)

then for the nonrelativistic and relativistic gases one has respectively $\gamma = \frac{2}{3}$ and $\gamma = \frac{1}{3}$, as will be seen in Section 8.3.

The total pressure of the crystal is

$$P = P_0 + \gamma \frac{U}{V} \qquad (4.4)$$

where P_0 is the pressure at absolute zero (where none of the Debye waves is excited). We have from (4.4)

$$\left(\frac{\partial P}{\partial T}\right)_V = \frac{\gamma}{V} C_V. \qquad (4.5)$$

Substituting in

$$\left(\frac{\partial P}{\partial T}\right)_V \left(\frac{\partial T}{\partial V}\right)_P \left(\frac{\partial V}{\partial P}\right)_T = -1,$$

one finds

$$\beta = \frac{1}{V}\left(\frac{\partial V}{\partial T}\right)_P = \frac{\gamma}{V} C_V K. \qquad (4.6)$$

This relation permits one to determine γ experimentally.

Equation (4.5) can also be applied to obtain an order of magnitude relation roughly applicable to polymorphic phase transitions in solids. We have by (4.5)

$$\left(\frac{\partial S}{\partial V}\right)_T = \frac{\gamma}{V} C_V.$$

This relation holds strictly for only a single phase. However one may try to apply it as an order of magnitude relation to obtain the entropy change between two phases [1]. With such a rough assumption one has for the slope of the phase line

$$\frac{dP}{dT} = \frac{\gamma}{V} C_V.$$

Putting in typical values $\gamma \cong 2$, $C_V = 3R$, and $V = 10$ cm^3, one finds

$$\frac{dP}{dT} \cong 50 \text{ atm/degree,}$$

which agrees approximately with the empirically determined typical slopes.

The Debye model is an enormous oversimplification, since it depends on only a single parameter, (Θ), which may be fixed by fitting the specific heat curve or by calculation from the elastic constants by equation (3.5). The only difference between a crystal of lead and of diamond, according to the model, is the value of Θ.

7.5 Electromagnetic Field

Electromagnetic radiation in thermodynamic equilibrium may also be described by the preceding calculation. Since the differential equations are again linear, the Hamiltonian is quadratic and may be exhibited as the sum of contributions from uncoupled harmonic oscillators. Consequently one again has by (3.1)

$$A = kT \int_0^\infty f(\omega)\, \rho(\omega)\, d\omega \tag{5.1}$$

where

$$f(\omega) = \ln 2 \sinh \frac{\hbar\omega}{2kT} \tag{5.1a}$$

$$\rho(\omega) = a\omega^2. \tag{5.1b}$$

There are two differences from the crystal. First, according to the usual electrodynamics there is no cutoff, $\bar{\omega}$; i.e., there is no smallest length or lattice structure to space. Second, a is slightly different since there are no longitudinal modes; it is now

$$a = 4\pi \left(\frac{2}{c^3}\right) \frac{V}{(2\pi)^3} \tag{5.2}$$

instead of (3.4a). We may calculate the total field energy at temperature T according to the equation

$$U = -T^2 \frac{\partial}{\partial T}\left(\frac{A}{T}\right)$$

$$= -kT^2 \int_0^\infty \frac{\partial f}{\partial T} \rho(\omega)\, d\omega$$

$$U = \frac{1}{2} \int_0^\infty (\hbar\omega)\left(\coth \frac{\hbar\omega}{2kT}\right) \rho(\omega)\, d\omega. \tag{5.3}$$

Hence

$$\frac{dU}{d\omega} = \frac{\hbar\omega}{2}\left(\coth \frac{\hbar\omega}{2kT}\right) \rho(\omega)$$

$$= \rho(\omega)\frac{\hbar\omega}{2} + \frac{\hbar\omega}{e^{\hbar\omega/kT} - 1}. \tag{5.4}$$

Except for the contribution of zero-point energy $\rho(\omega)\cdot\hbar\omega/2$, equation (5.4) is Planck's law. By integrating over all frequencies (but not counting the zero-point energy), one obtains the Stefan-Boltzmann law:

$$U = \int_0^\infty \frac{\hbar\omega\, \rho(\omega)\, d\omega}{e^{\hbar\omega/kT} - 1}$$

$$= a\left(\frac{kT}{\hbar}\right)^4 \hbar \int_0^\infty \frac{x^3\, dx}{e^x - 1} \tag{5.5}$$

where

$$x = \frac{\hbar\omega}{kT}.$$

Hence

$$U = \sigma T^4 V \tag{5.6}$$

where

$$\sigma = \frac{48\alpha\pi k^4}{h^3 c^3} \tag{5.6a}$$

and

$$\alpha = \frac{1}{6}\int_0^\infty \frac{x^3\, dx}{e^x - 1}$$

$$= 1 + \frac{1}{2^4} + \frac{1}{3^4} + \cdots = 1.0823.$$

Since the energy content depends on T^4, the specific heat varies as T^3, just as for a crystal at low temperature, and for the same reason. The formal expression for the energy is essentially the same in the two cases; the upper limit of the integral Θ/T becomes infinite for the crystal when $T \longrightarrow 0$ and is always infinite for radiation since $\Theta = \infty$ in this case.

In order to determine the thermal equation of state, it is convenient to exhibit the volume and temperature dependence of A explicitly:

$$A = -cVT^4 \tag{5.7}$$

by (5.1) where c is a constant. Then

$$P = -\left(\frac{\partial A}{\partial V}\right)_T = -\frac{A}{V}. \tag{5.8}$$

Hence

$$G(T, P) = A + PV = 0. \tag{5.9}$$

This is the thermal equation of state. (We shall see later that the vanishing of G means that the number of particles is not conserved.)

Notice also that the energy density is

$$\frac{U}{V} = \frac{1}{V}\left(A - T\frac{\partial A}{\partial T}\right)$$

$$= -\frac{3A}{V} = 3P$$

or

$$P = \frac{1}{3}\frac{U}{V}. \tag{5.10}$$

In this case the ratio of the pressure to the energy density is $1/3$, which is appropriate for relativistic particles with negligible or vanishing rest mass, such as photons. (Compare with equation (4.3) of this chapter.)

7.6   Paramagnetism

An ideal paramagnetic system is composed of noninteracting magnets. We shall now consider the important example of a crystal containing a sublattice of magnetically active ions that are far apart and separated by magnetically inactive ions. Salts of the rare earths, when they are made magnetically dilute by separating the magnetically active rare earth ions by large numbers of inert atoms, are very good illustrations of the theory because the individual rare earth ions behave very much like free ions [2]. Of course there is always a weak interaction between the magnetically active ions, no matter how dilute the salt is. This interaction remains negligible as long as it is small compared to kT, but at sufficiently low temperatures, this weak interaction will induce a phase transition in which the system of magnetic moments becomes aligned either ferromagnetically or antiferromagnetically.

In the paramagnetic phase the Hamiltonian ($\mathcal{3C}$) is

$$\mathcal{3C} = \sum_i \mathcal{3C}_i \tag{6.1}$$

where $\mathcal{3C}_i$ is the energy of interaction of a single magnetic ion with the external magnetic field. Hence the previous formulas for an ideal system are valid, namely,

$$Q = \Pi\, Q_i$$

$$A = \sum_i A_i.$$

The change in the energy of an atom caused by the application of a field, H, is described by the following formula for the Zeeman effect:

$$E(M) = E_0 - g\beta H M \tag{6.2}$$

to terms of the first order in H. According to this formula an atomic level, E_0, described by quantum numbers (L, S, J), and denoted by $^{2S+1}L_J$ is split into $2J + 1$ levels by the magnetic field, H. Here β is the Bohr magneton:

$$\beta = \frac{e\hbar}{2mc} \tag{6.3}$$

and g is the Lande factor, which for free ions is

$$g = \frac{3}{2} + \frac{S(S+1) - L(L+1)}{2J(J+1)}. \tag{6.4}$$

For free ions we have $g = 1$ for $S = 0$, and $g = 2$ for $L = 0$. The value of g of course depends on the wave function and in solids it will deviate from (6.4).

The magnetic partition function for the set of Zeeman levels is

$$Q = \sum_{-J}^{+J} e^{-(E_0 - g\beta HM)/kT}$$

$$= e^{-E_0/kT} \sum_{-J}^{+J} e^{+g\beta HM/kT}$$

$$= e^{-E_0/kT} \sum_{-J}^{+J} x^M \tag{6.5}$$

where

$$x = e^{-g\beta H/kT}$$

$$\sum_{-J}^{+J} x^M = \frac{x^{J+\frac{1}{2}} - x^{-J-\frac{1}{2}}}{x^{1/2} - x^{-1/2}}.$$

In the following considerations we may put $E_0 = 0$. The free energy calculated from Q is for N atoms

$$A = - kNT \left\{ \ln \sinh \left(J + \frac{1}{2} \right) \frac{g\beta H}{kT} - \ln \sinh \frac{1}{2} \frac{g\beta H}{kT} \right\}. \tag{6.6}$$

Consider next the magnetic moment. For the particular state (α) with energy E_α, the magnetic moment is

$$M_\alpha = - \frac{\partial E_\alpha}{\partial H}. \tag{6.7}$$

If only the temperature is given, the atom has in general a nonnegligible probability of existing in many different quantum states and the average magnetic moment is

$$M = \langle M_\alpha \rangle = \frac{\sum_\alpha e^{-E_\alpha/kT}\left(-\dfrac{\partial E_\alpha}{\partial H}\right)}{\sum_\alpha e^{-E_\alpha/kT}}$$

$$= kT \frac{\partial}{\partial H} \ln Q$$

$$= -\left(\frac{\partial A}{\partial H}\right)_T. \tag{6.8}$$

Hence the magnetic moment may also be calculated directly from the free energy.

Let us write

$$A = - RT \, \mathscr{L}_J(\theta) \tag{6.9}$$

where

$$\mathscr{L}_J(\theta) = \ln \frac{\sinh\left(J + \dfrac{1}{2}\right)\theta}{\sinh \dfrac{1}{2}\theta} \tag{6.9a}$$

and

$$\theta = \frac{g\beta H}{kT}. \tag{6.9b}$$

It follows that

$$M = RT \left(\frac{\partial \theta}{\partial H}\right)_T \mathscr{L}'(\theta)$$

$$= Ng\beta \, \mathscr{L}'_J(\theta) \tag{6.10}$$

where

$$\mathscr{L}'_J(\theta) = \left(J + \frac{1}{2}\right) \coth\left(J + \frac{1}{2}\right)\theta - \frac{1}{2}\coth\frac{\theta}{2}, \tag{6.10a}$$

or

$$M = Ng\beta J B_J(\theta) \tag{6.11}$$

where

$$J B_J(\theta) = \mathscr{L}'_J(\theta). \tag{6.11a}$$

B_J is called the Brillouin function. For weak fields we have $\theta \cong 0$, $\coth \theta = \frac{1}{\theta} + \frac{\theta}{3}$. Hence

$$\mathcal{L}'(\theta) = \left(J + \frac{1}{2}\right)\left[\frac{1}{\left(J + \frac{1}{2}\right)\theta} + \left(J + \frac{1}{2}\right)\frac{\theta}{3}\right] - \frac{1}{2}\left[\frac{2}{\theta} + \frac{\theta}{6}\right]$$

$$= J(J + 1)\frac{\theta}{3}. \tag{6.12}$$

Hence the magnetic susceptibility, χ, is

$$\chi = \lim_{H \to 0} \left(\frac{M}{H}\right) = \frac{1}{3}\frac{N^2 g^2 \beta^2 J(J + 1)}{kT}. \tag{6.13}$$

This may be compared with the classical formula for the susceptibility

$$\chi = \frac{N\mu^2}{3kT}. \tag{6.14}$$

These agree if $\mu^2 = J(J + 1)g^2\beta^2$. The factor $1/3$ in both the classical and quantum formulas results from an averaging over orientations in space.

At the other extreme of very high field strengths, $M(H)$ approaches the saturation value $M_\infty (= Ng\beta)$, since

$$M(H) = M_\infty B_J(\theta)$$

and

$$\lim_{H \to \infty} B_J(\theta) = 1.$$

The caloric properties may be calculated from the equations

$$S = -\left(\frac{\partial A}{\partial T}\right)_H$$

$$U = -T^2 \frac{\partial}{\partial T}\left(\frac{A}{T}\right)$$

$$C_H = -\frac{\partial}{\partial T} T^2 \frac{\partial}{\partial T}\left(\frac{A}{T}\right).$$

We find

$$\frac{S}{R} = \mathcal{L}(\theta) - J\theta B_J(\theta) \tag{6.15}$$

$$C_H = \theta^2 \frac{d^2\mathcal{L}}{d\theta^2}. \tag{6.16}$$

All the preceding caloric equations imply that the function, U, does not contain the field energy and therefore correspond to the form of the first law as given in note [11] of Chapter 3.

7.7   Adiabatic Demagnetization and the Third Law

The preceding results may be applied to make an analysis of adiabatic demagnetization. In particular note that the entropy is a function of the parameter:

$$\theta = \frac{g\beta H}{kT}.$$

Let us now consider for simplicity a doubly degenerate level which splits into a doublet in a magnetic field. Let us assume that the degeneracy is not perfect and that the two ground states have slightly different energies on account of perturbations by the rest of the crystal.

As mentioned before this perturbation may be caused by the Stark splitting produced by the crystalline electric field. Denote the width of the doublet at zero field by δ. Then the entropy is determined by

$$\theta = \frac{\delta + g\beta H}{kT}, \tag{7.1}$$

since the entropy depends only on the total width of the doublet.

Let bc in Figure 7.2 denote an adiabatic demagnetization. Then

$$S_c = S_b. \tag{7.2}$$

This implies, since $S = S(\theta)$,

$$\theta_c = \theta_b \tag{7.3}$$

or

$$\frac{\delta}{kT_c} = \frac{g\beta H + \delta}{kT_b} \cong \frac{g\beta H}{kT_b} \tag{7.4}$$

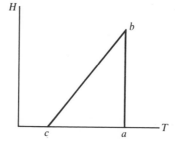

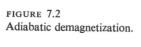

FIGURE 7.2
Adiabatic demagnetization.

where we have assumed $g\beta H \gg \delta$. Hence

$$\frac{T_c}{T_b} = \frac{\delta}{g\beta H}. \tag{7.5}$$

From this formula one may estimate the final temperature. According to this result the final temperatures achievable are limited by interatomic forces which prevent δ from having a vanishingly small value. If the splitting caused by an interatomic interaction is expressed as $\delta = kT^*$ then one may say that the adiabatic demagnetization runs into difficulties at $T = T^*$.

Also if δ is very small (if interatomic interactions are very weak), then a long time is required to establish thermal equilibrium.

In this situation the Stark splitting provides a mechanism for implementing the third law since we would have, according to equation (7.5),

$$T_c = 0$$

if $\delta = 0$. That is, if the ground state were degenerate in the absence of the magnetic field, absolute zero would be attainable. Nernst's statement would of course also be violated since the magnetic field would lift this degeneracy and therefore we would have

$$\lim_{T \to 0} [S(H, T) - S(0, T)] \neq 0.$$

The general conjecture expressed by the third law is that some mechanism always comes into play to remove the degeneracy of the ground state and establish the third law in Planck's form

$$S(T = 0) = 0.$$

In the case of a paramagnetic system an obvious mechanism is provided by the always present (if weak) interaction between the magnetic ions. As the temperature is lowered, this interaction will always induce a Curie transition in agreement with the thermodynamic result that Curie's law cannot hold all the way to absolute zero [see (13.7) of Chapter 4]. If a system of nuclear spins is used, instead of an electronic system, the same general remarks hold, although one can then reach temperatures of the order of 10^{-6} K instead of 10^{-3} K before the method breaks down.

Another important general mechanism for lifting degeneracy is the so-called Jahn-Teller effect which distorts a symmetric configuration of nuclei in order to remove electronic degeneracy of the ground state [3].

Finally the significance of superfluidity and superconductivity as manifestations of the third law has been mentioned repeatedly. In those examples there is a weak residual interaction between the helium atoms in the one case and the electrons in the other; in both cases the effect of the weak interaction and statistics is to lift the degeneracy of the unperturbed ground state by splitting off a new ground state and separating it from the rest of the spectrum by an effective energy gap.

7.8   Negative Temperatures

According to the third law, absolute zero is not attainable but there is no similar theoretical reason that excludes states of negative temperature. It is true that these situations are not permitted by classical mechanics; for if $T < 0$ then the Boltzmann factor favors systems of high kinetic energy which fly apart. On the other hand it is possible to prepare quantum systems that are essentially isolated and have a bounded energy spectrum; these systems may exhibit negative temperatures.

Consider for example a system of nuclear spins that interact among themselves by spin-spin magnetic coupling. The spins may come to equilibrium in a time that is short compared to the time required to establish equilibrium with the translational motions of the nuclei or with the electronic environment. In these situations the spin system may exist for some time at a negative temperature while the surrounding lattice is at positive temperature.

A system of nuclear systems is almost a perfect example of a paramagnetic system since the interaction between the spins is so weak. The previous formulas of section 6.1 are then valid. According to these expressions the entropy is a symmetrical curve centered at the unperturbed level. That is, the partition function is

$$Q = \sum_{-J}^{+J} e^{-(E_0 + \Delta_M)/kT} = e^{-E_0/kT} \, \bar{Q}$$

$$Q = \sum_{-J}^{+J} e^{-\Delta_M/kT} \tag{8.1}$$

where Δ_M is the magnetic splitting. The following substitution

$$\Delta_M \longrightarrow -\Delta_M$$
$$T \longrightarrow -T \tag{8.2}$$

leaves the reduced partition function ($\bar{Q}$) and the entropy unchanged but changes the sign of the free energy:

$$\bar{Q} \longrightarrow \bar{Q}$$
$$\bar{A} \longrightarrow -\bar{A}, \tag{8.3}$$

and therefore

$$\bar{U} \longrightarrow -\bar{U}$$
$$S \longrightarrow S. \tag{8.4}$$

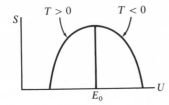

FIGURE 7.3
Entropy as function of internal energy.

The corresponding curve is sketched in Figure 7.3. The entropy, which measures the number of states, vanishes at U_{max} and U_{min} where all the spins are aligned and there is only a single state possible. We have

$$\frac{1}{T} = \left(\frac{\partial S}{\partial U}\right)_H. \tag{8.5}$$

Hence if $U < E_0$, then $T > 0$; but if $U > E_0$, then $T < 0$. (In the normal classical situation the magnitude of S, or the amount of phase space, increases as the energy is increased.)

A negative temperature means that the Boltzmann factor favors the excited states; therefore if the temperature of a system is changed from $+T$ to $-T$, its energy is increased. If two systems, with temperatures $+T$ and $-T$, are put into interaction with each other so that they reach a common final temperature, then energy must flow from $-T$ to T in order to conserve energy, i.e., negative temperatures are "hotter".

The hottest temperature is therefore $T = 0 - \epsilon$ and the coldest $T = 0 + \epsilon$. The temperature scale from cold to hot runs as follows:

$$0° \cdots + \infty - \infty \cdots - 0°.$$

A better measure than T would be $-\dfrac{1}{T} = \lambda$ which runs

$$- \infty° \cdots - 0° + 0° \cdots + \infty°.$$

Formally the possibility of realizing states of negative temperature arises because the energy spectrum has an upper limit. That is

$$Q(\lambda) = \int_{-\infty}^{+\infty} e^{\lambda E} \rho(E)\, dE. \tag{8.6}$$

In any case

$$\rho(E) = 0 \quad \text{if} \quad E < 0 \tag{8.6a}$$

but for systems that may exhibit negative temperatures, one has in addition

$$\rho(E) = 0 \quad \text{if} \quad E > E_{max}. \tag{8.6b}$$

According to the Titchmarsh theorem [4], it follows from (8.6a) that $Q(\lambda)$ is an analytic function in the right hand T-plane (left-hand λ-plane) provided that $\rho(E)$ is quadratically integrable.

According to the same theorem, if (8.6b) also holds, then $\exp(-\lambda E_{\max})Q(\lambda)$ is an analytic function in the left-hand T-plane as well. Thus the condition (8.6b) permits the continuation of the partition function and thermodynamic functions into regions of negative temperature.

Notes and References

1. J. C. Slater, *Introduction to Chemical Physics*, p. 220 (McGraw-Hill, New York, 1939). Here one is roughly assuming that ΔS depends only on ΔV and does not depend on whether the change takes place in one phase or between two phases.
2. The reason for this is that the electrons contributing to the magnetic moment are strongly shielded, with the result that the state of these electrons can be worked out by starting with the wave function of a free ion and calculating the perturbation which is caused by the electric field coming from the crystalline environment (the crystalline field).
3. H. A. Jahn and E. Teller, *Proc. Roy. Soc.* (London) 161: 220 (1937).
4. E. C. Titchmarsh, *Introduction to the Theory of the Fourier Integral*, Chapter V (Oxford Univ. Press, Oxford, 1937).

   In the present context the theorem takes the following forms: Define $\tilde{Q}(\zeta)$ to be the Fourier transform of $\rho(E)$:

$$\tilde{Q}(\zeta) = \int_{-\infty}^{+\infty} e^{i\zeta E}\rho(E)\,dE,$$

and if

$$\int_{-\infty}^{+\infty} \rho(E)^2\,dE < \infty$$

$$\rho(E) = 0 \quad \text{if} \quad E < 0,$$

then $\tilde{Q}(\zeta)$ is the boundary value of a function $\tilde{Q}(\zeta_1 + i\zeta_2)$ which is analytic in the upper half plane ($\zeta_2 > 0$). (The boundary is $\zeta_2 \longrightarrow 0+$.) But if $\zeta = -i\lambda$, then

$$\tilde{Q}(\zeta) = \int_{-\infty}^{+\infty} e^{\lambda E}\rho(E)\,dE$$

$$= Q(\lambda),$$

and therefore $Q(\lambda)$ is the boundary value of a function $Q(\lambda_1 + i\lambda_2)$ which is analytic in the left-hand plane ($\lambda_1 < 0$) or the right-hand temperature plane. (The boundary is $\lambda_1 \longrightarrow 0-$.) Next consider

$$Q(\lambda)e^{-\lambda E_{\max}} = \int_{-\infty}^{+\infty} e^{\lambda(E - E_{\max})}\rho(E)\,dE.$$

Let

$$\rho(E) = \sigma(x)$$

where

$$x = E - E_{\max}.$$

Then

$$Q(\lambda)e^{-\lambda E_{\max}} = \int_{-\infty}^{+\infty} e^{\lambda x}\sigma(x)\, dx$$

where

$$\sigma(x) = 0 \quad \text{if} \quad x > 0.$$

Hence $Q(\lambda) \exp(-\lambda E_{\max})$ is analytic in the left-hand temperature-plane. (The transform of a function that vanishes outside a finite interval is analytic in the entire plane, except perhaps at infinity, of the transformed variable. If it vanishes everywhere except at a point, it is a δ-function, and its transform is a constant, which is analytic everywhere.)

8

Thermodynamics of Ideal Quantum Gases

8.1 Quantum Statistics of Ideal Gases

The thermodynamics of a physical system depends in an important way on the spin of the individual particles. This dependence does not come about primarily through the spin dependent forces, since these are very weak, but through the following connection between the spin of the individual particles and the symmetry of the wave function: namely, if the spin is an even (odd) multiple of $\hbar/2$, then the wave function is symmetric (antisymmetric). This connection between spin and statistics is imposed by the demands of special relativity but holds universally, even in situations that would ordinarily be judged nonrelativistic.

In the case of noninteracting particles, the Hamiltonian is additive:

$$H = \sum_i H_i. \tag{1.1}$$

Let the wave function belonging to the partial Hamiltonian H_i be $u(x_i; a_i)$ where a_i is the set of quantum numbers describing the ith particle. Then we have the following results depending on the nature of the particles:

$$\Psi(x_1, \ldots, x_N; a_1, \ldots, a_N) = \prod_{i=1}^{N} u(x_i; a_i) \qquad \text{distinguishable particles} \tag{1.2}$$

$$\Psi_S = \sum_P P\Psi(x_1, \ldots, x_N; a_1, \ldots, a_N) \quad \text{identical particles}$$

$$\text{spin} = n\hbar \quad (1.3)$$

$$\Psi_A = \sum_P \epsilon_P P\Psi(x_1, \ldots, x_N; a_1, \ldots, a_N) \quad \text{identical particles}$$

$$\text{spin} = \left(n + \frac{1}{2}\right)\hbar. \quad (1.4)$$

Here P is an operator that permutes the quantum numbers $(a_1, \ldots, a_N)$, and the sum runs over all $(N!)$ permutations. Each permutation operator may be written as a product of transpositions; if the number of transpositions is even (odd), the permutation is called even (odd). The coefficient is $\epsilon_P = +1$ for even permutations and $\epsilon_P = -1$ for odd permutations, i.e., (1.3) is a permanent and (1.4) is a determinant. For a given set of quantum numbers $a_1, \ldots, a_N$, there is only one symmetrical function Ψ_S and only one antisymmetric function Ψ_A. If the particles are distinguishable, however, each permuted product represents a different possible state. Therefore the identity of the particles reduces the number of allowed states by $N!$.

Particles with symmetric (antisymmetric) wave functions are said to obey Einstein-Bose (Fermi-Dirac) statistics. In the Einstein-Bose (E.B.) case, it is possible for any number of the sets $\{a_i\}$ to be equal. In the Fermi-Dirac (F.D.) case the wave function vanishes if two of the sets $\{a_i\}$ agree; this is the Pauli exclusion principle.

It follows that for identical particles, a conceivable state, if it exists at all, is completely described by the population numbers $n_1, n_2, \ldots$ of the states $a_1, a_2, \ldots$. It is meaningless to say which particles are in the separate states; one can only say how many particles are in the separate states. In the F.D. case, $n_i = 0, 1$.

We shall find that the quantum formulas, except for the entropy, approach the classical expressions under conditions of low density and high temperature (weak degeneracy). The limiting form of the entropy expression fortunately does not agree with the classical formula, since the latter has a fundamental defect that shows up in the so-called Gibbs paradox.

In the other extreme (strong degeneracy), the quantum gases show properties that are completely nonclassical. For example the first approximation to a superconducting system is provided by a completely degenerate F.D. gas, and the first approximation to a superfluid system is provided by a completely degenerate E.B. gas; however, there is in both cases a very weak, but essential, interaction between the particles which produces the superstates.

8.2   Partition Function of Ideal Gases

We consider a gas composed of identical noninteracting particles. The total energy of the gas in the state (α) is

$$E(\alpha) = E(a_1) + E(a_2) + \cdots \tag{2.1}$$

where a_i is the set of quantum numbers belonging to the ith particle and α is the complete set of quantum numbers for the entire gas: $\alpha = (a_1, a_2, \ldots)$. The partition function [1] is

$$Q = \sum_\alpha e^{-E(\alpha)/kT} \tag{2.2}$$

where the sum is extended over all states. Since the particles are identical each has the same spectrum, and the quantum state α may alternatively be specified by a set of integers

$$\alpha = (n_1, n_2, \ldots) \tag{2.3}$$

where $n_1, n_2, \ldots$ are the populations of the different one-particle states. This last equation is a postulate about the system. It tells us that a state is not changed if any permutation of the particles is made that does not alter the population numbers $(n_1, n_2, \ldots)$. If the individual particles are distinguishable, such an assumption is not valid.

For the total energy one now has

$$E(\alpha) = n_1\epsilon_1 + n_2\epsilon_2 + \cdots \tag{2.4}$$

and

$$Q_n = \sum_{n_1+n_2+\cdots=n} \sum \sum \cdots \exp\left[-(n_1\epsilon_1 + n_2\epsilon_2 + \cdots)/kT\right] \tag{2.5}$$

where n is the total number of particles. Then

$$Q_n(x_1, x_2, \ldots) = \sum_{n_1+n_2+\cdots=n} x_1^{n_1} x_2^{n_2} \cdots \tag{2.6}$$

where

$$x_i = \exp(-\epsilon_i/kT). \tag{2.6a}$$

One now has to distinguish between fermions, which obey the Fermi-Dirac statistics, and bosons, which obey the Einstein-Bose statistics. Fermions have antisymmetric wave functions; bosons have symmetric wave functions. For fermions,

$$n_k = 0, 1. \tag{2.7}$$

For bosons, n_k is not restricted. The limitation on fermions is the Pauli principle: the statement that the antisymmetric wave function vanishes if two particles are in the same state. The partition function is easily evaluated if $n = \infty$. Then

$$Q_\infty = \sum_{n_1} x_1{}^{n_1} \sum_{n_2} x_2{}^{n_2} \cdots$$

$$= (1 + x_1)(1 + x_2) \cdots \qquad \text{for Fermions}$$

$$= (1 - x_1)^{-1}(1 - x_2)^{-1} \cdots \qquad \text{for Bosons}$$

or

$$Q_\infty(x_1, x_2, \ldots) = \Pi_i(1 \pm x_i)^{\pm 1}. \tag{2.8}$$

Now consider

$$Q_\infty(\zeta x_1, \zeta x_2, \ldots) = \sum_{n_1, n_2, \ldots} (\zeta x_1)^{n_1}(\zeta x_2)^{n_2} \cdots$$

$$= \sum_{n=0}^{\infty} \zeta^n \left(\sum_{n_1 + n_2 + \cdots = n} x_1{}^{n_1} x_2{}^{n_2} \cdots \right)$$

$$= \sum_{n=0}^{\infty} \zeta^n Q_n(x_1, x_2, \ldots). \tag{2.9}$$

It follows that

$$Q_n(x_1, x_2, \ldots) = \frac{1}{2\pi i} \oint \zeta^{-n-1} Q_\infty(\zeta x_1, \ldots) \, d\zeta. \tag{2.10}$$

This result is an exact formal representation of $Q_n(x_1, x_2, \ldots)$ since we already know $Q_\infty(\zeta x_1, \zeta x_2, \ldots)$; the contour integral projects out the part of Q_∞ which is homogeneous of degree n. For large n one sees that $\zeta^{-n-1} Q_n(\zeta x_1, \zeta x_2, \ldots)$ has a very sharp minimum on the real axis. Let the position of this minimum be $\bar{\zeta}$. Since the integrand is analytic, the minimum in the real direction is accompanied by a maximum in the imaginary direction, which is also very sharp. By choosing the contour to be a circle through $\bar{\zeta}$, we may approximate the integral by the method of steepest descents, or even more crudely by

$$Q_n \cong \frac{1}{2\pi i} [\zeta^{-n-1} Q_\infty(\zeta)]_{\bar{\zeta}} \cdot (i\Delta) \tag{2.11}$$

where Δ is the effective width of the maximum [1].
Then

$$A_n \cong -kT \ln [\bar{\zeta}^{-n} Q_\infty(\bar{\zeta})]$$

$$= kT [n \ln \bar{\zeta} - \ln Q_\infty(\bar{\zeta})] \tag{2.12}$$

where terms independent of n have been dropped Since $\bar{\zeta}$ is determined by the minimum in A, we may also write

$$A_n = nkT \ln \zeta + A_\infty(\zeta) \tag{2.13}$$

subject to the side condition:

$$\frac{\partial A_n}{\partial \zeta} = 0 \tag{2.13a}$$

where

$$A_\infty(\zeta) = -kT \ln Q_\infty(\zeta)$$
$$= \mp kT \sum_s \ln (1 \pm \zeta x_s). \tag{2.13b}$$

Hence we also have

$$A_n = kT \left[n \ln \zeta \mp \sum_s \ln (1 \pm \zeta x_s) \right] \tag{2.13c}$$

still subject to (2.13a). Notice that Δ, the width of the maximum, has now completely dropped out. Various thermodynamic functions may be computed from A_n in the usual way. In addition, the expectation value of the number of particles in the ith single particle state may be determined as follows:

$$\langle n_i \rangle = \frac{\sum n_i e^{-E/kT}}{\sum e^{-E/kT}} \tag{2.14}$$

where the sum is extended over all states of the complete system, and where

$$E = \sum_i n_i \epsilon_i.$$

Hence

$$\langle n_i \rangle = \frac{\sum \frac{\partial E}{\partial \epsilon_i} e^{-E/kT}}{\sum e^{-E/kT}}$$

$$= -kT \frac{\partial}{\partial \epsilon_i} \ln \sum e^{-E/kT}$$

or

$$\langle n_i \rangle = \frac{\partial A_n}{\partial \epsilon_i}. \tag{2.15}$$

Applying this formula to (2.13c) we obtain

$$\langle n_i \rangle = \frac{\zeta x_i}{1 \pm \zeta x_i}$$

$$= \frac{1}{\zeta^{-1} x_i^{-1} \pm 1}. \tag{2.16}$$

The expression for $\langle n_i \rangle$ as well as the original formula for A contains the undetermined parameter ζ. This parameter is fixed by the side condition

$$\frac{\partial A_n}{\partial \zeta} = 0$$

or

$$n = \sum_i \frac{1}{\zeta^{-1} x_i^{-1} \pm 1},$$

which is the same as

$$n = \sum_i \langle n_i \rangle. \tag{2.17}$$

The equation fixing ζ thus has a simple interpretation according to (2.17). Given the total number of particles and the various Boltzmann factors, (2.17) is an implicit equation for ζ.

It is finally convenient to introduce degeneracy factors g_i. Then

$$A_n = nkT \ln \zeta \mp kT \sum_i g_i \ln (1 \pm \zeta x_i) \tag{2.18}$$

$$n = \sum_i \frac{1}{\zeta^{-1} x_i^{-1} \pm 1} g_i \tag{2.19}$$

where g_i is the number of single particle states corresponding to the single particle energy ϵ_i. Since the single particle energy levels are very closely spaced, their spectrum may be approximated by a continuum and the degeneracy g_i may be replaced by a continuous function $g(\epsilon)$, such that $g(\epsilon) \, d\epsilon$ is the number of states in the interval $d\epsilon$. The explicit formula for $g(\epsilon)$ may be determined from the rule that there are $2s + 1$ states for each cell of volume h^3 in phase space, where s is the spin [2]. Then

$$g(\epsilon) \, d\epsilon = (2s + 1)V \frac{4\pi p^2 \, dp}{h^3}$$

$$= \mathbf{V} \frac{4\pi p^2 \, dp}{h^3} \tag{2.20}$$

where $\mathbf{V}$ is the volume of gas multiplied by $(2s + 1)$ and $4\pi p^2 \, dp$ is the volume of a shell in momentum space. Now substituting in (2.18) and suppressing the subscript on A_n, we have

$$\frac{A}{kT} = n \ln \zeta \mp \int_0^\infty \frac{4\pi \mathbf{V} p^2 \, dp}{h^3} \ln (1 \pm \zeta x). \tag{2.21}$$

Put

$$y^2 = \frac{p^2}{2mkT}.$$

Then

$$\frac{A}{RT} = \ln \zeta \mp \frac{1}{\theta} F_2(\zeta) \qquad (2.22)$$

where

$$F_2(\zeta) = 4\pi \int_0^\infty \ln (1 \pm \zeta e^{-v^2}) y^2 \, dy, \qquad (2.22a)$$

and

$$\theta = \frac{nh^3}{V(2mkT)^{3/2}} \qquad (2.22b)$$

is a dimensionless parameter distinguishing one gas from another. $F_2(\zeta)$ may also be written after an integration by parts as

$$F_2(\zeta) = \pm \frac{8\pi}{3} \int_0^\infty \frac{y^4 \, dy}{\zeta^{-1}e^{y^2} \pm 1}. \qquad (2.22c)$$

Likewise the condition for ζ is, by (2.19),

$$n = \int_0^\infty \frac{1}{\zeta^{-1}e^{y^2} \pm 1} \frac{4\pi V p^2 \, dp}{h^3}$$

or

$$\theta = F_1(\zeta) \qquad (2.23)$$

where

$$F_1(\zeta) = 4\pi \int_0^\infty \frac{y^2 \, dy}{\zeta^{-1}e^{y^2} \pm 1}. \qquad (2.23a)$$

Equation (2.23) determines ζ if θ is assigned. Together ζ and θ fix A. Hence A and therefore the thermodynamics depends only on the single dimensionless parameter

$$\theta = \frac{nh^3}{V(2mkT)^{3/2}}. \qquad (2.24)$$

Note that $(2mkT)^{1/2}$ is a characteristic momentum, say $\bar{p}$, and that $h/\bar{p} = \bar{\lambda}$ is a wave length associated with $\bar{p}$. We may write

$$\theta = \frac{\bar{\lambda}^3}{(V/n)}. \qquad (2.24a)$$

Hence the parameter θ measures the ratio of $\bar{\lambda}^3$ to the volume which may be apportioned to a single particle. This ratio is large at low temperatures when the wave length becomes long and at high densities when V/n becomes small. When θ is large, the quantum mechanical statistics differ greatly from the Maxwell-Boltzmann limit; when θ is small, quantum mechanical effects are small.

8.3   Equations of State

The thermal equation of state is

$$P = -\left(\frac{\partial A}{\partial V}\right)_T = -\left(\frac{\partial A}{\partial \zeta}\right)_\theta \left(\frac{\partial \zeta}{\partial V}\right)_T - \left(\frac{\partial A}{\partial \theta^{-1}}\right)_\zeta \left(\frac{\partial \theta^{-1}}{\partial V}\right)_T$$

or

$$P = \pm F_2(\zeta) RT \frac{\partial}{\partial V} (\theta^{-1}). \tag{3.1}$$

Again the fact that $\partial A / \partial \zeta = 0$ makes the first term vanish. Hence

$$\frac{PV}{RT} = \pm \frac{F_2(\zeta)}{\theta}. \tag{3.2}$$

The caloric equation is

$$U = -T^2 \frac{\partial}{\partial T} \left(\frac{A}{T}\right). \tag{3.3}$$

Hence

$$\frac{U}{R} = \pm T^2 \frac{\partial}{\partial T} \left(\frac{F_2(\zeta)}{\theta}\right)$$

$$U = \pm \frac{3}{2} RT \frac{F_2(\zeta)}{\theta}. \tag{3.4}$$

The two equations (3.2) and (3.4) lead to the result

$$PV = \frac{2}{3} U. \tag{3.5}$$

This exact result is independent of the statistics and states that the pressure is 2/3 (energy density). It results from substituting the nonrelativistic connection between energy and momentum in A. If the relativistic relation $\epsilon = pc$ had been used, we would have obtained

$$PV = \frac{1}{3} U. \tag{3.6}$$

By combining the equation (2.22) for A with (3.2) above, we obtain an explicit formula for

$$\ln \zeta = \frac{A}{RT} \pm \frac{F_2(\zeta)}{\theta}$$

$$= \frac{A + PV}{RT} = \frac{G}{RT} \tag{3.7}$$

or

$$\zeta = e^{G/RT}. \tag{3.7a}$$

This result is also exact. The distribution formula (2.16) in terms of G instead of ζ reads

$$\langle n_s \rangle = \frac{1}{e^{(\epsilon - \epsilon_g)/kT} \pm 1} \tag{3.8}$$

where ϵ_g is the Gibbs energy per particle.

Entropy

By equations (3.5) and (3.7) the entropy is

$$\frac{S}{R} = \frac{U - A}{RT}$$

$$\frac{S}{R} = \frac{5}{2}\frac{PV}{RT} - \ln \zeta. \tag{3.9}$$

8.4   Weak Degeneracy

In this limit, $\theta \ll 1$ and ζ is also small. Therefore, by (2.23), for both the Fermi-Dirac and Einstein-Bose statistics,

$$\theta \cong 4\pi\zeta \int_0^\infty e^{-y^2} y^2 \, dy \tag{4.1}$$

and by (2.24)

$$\zeta \cong \frac{nh^3}{(2\pi mkT)^{3/2}} \frac{1}{2s + 1} \frac{1}{V}. \tag{4.2}$$

We also have by (3.2) and (2.23)

$$\frac{PV}{RT} = \pm \frac{F_2(\zeta)}{F_1(\zeta)} \tag{4.3}$$

When θ and ζ are small, by (2.22a) and (2.23a)

$$\frac{F_2(\zeta)}{F_1(\zeta)} = \frac{\pm \int_0^\infty \zeta e^{-v^2 y^2}\, dy}{\int_0^\infty \zeta e^{-v^2 y^2}\, dy} = \pm 1. \tag{4.4}$$

Hence by (4.3)

$$\lim_{\theta \to 0} \frac{PV}{RT} = 1. \tag{4.5}$$

The exact caloric equation of state remains the same:

$$P = \frac{2}{3}\frac{U}{V}. \tag{4.6}$$

Thus F.D. and E.B. gases approach the same weakly degenerate limit. Both classical equations of state are recovered; however the formula for the entropy differs from the classical result. We find by (3.9) and (4.2)

$$\frac{S}{k} = n\left[\frac{3}{2}\ln T + \ln\frac{V}{n} + \frac{s_0}{k}\right] \tag{4.7}$$

where

$$\frac{s_0}{k} = \frac{5}{2} + \ln\frac{(2\pi m k)^{3/2}(2s+1)}{h^3}. \tag{4.7a}$$

Equation (4.7) is the Sackur-Tetrode formula for the entropy of an ideal gas. It has the correct dependence on n and V in contrast to the classical expression for which the second term is simply

$$n \ln V. \tag{4.8}$$

As a result of this term, the classical expression incorrectly predicts an increase in entropy due to self-mixing. If the gases A and B in Figure 8.1 are different and the partition ab is removed, then there will be an increase in entropy due to the mixing of A and B. However if the gases A and B are the same, there can be no observable effect when ab is removed. This result is correctly described by the quantum expression for the entropy but not by the classical formula, since the term $n \ln V$ implies an increase in entropy of $n \ln 2$ (See section 2.8).

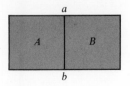

FIGURE 8.1
Diagram illustrating Gibbs paradox.

The quantum theory not only removes this classical paradox but also permits the completion of chemical thermodynamics, since it gives the expression for S_0/k. The equilibrium constant in the law of mass action may then be computed as follows.

The condition for chemical equilibrium is, by (2.3) of Chapter 5,

$$\sum_j \nu_j \mu_j = 0. \tag{4.9}$$

By (3.7) and (4.2) this condition may be put in the form

$$\sum_j \nu_j \ln P_j = \ln K(T) \tag{4.10}$$

where

$$\ln K(T) = \sum_j \nu_j \left[\frac{5}{2} \ln T - \ln \frac{h^3}{(2s+1)(2\pi mk)^{3/2}k} \right]_j. \tag{4.10a}$$

The constant of integration appearing in (3.6) of Chapter 5 has now been evaluated. In (4.10a) the degeneracy and the mass may depend on j.

8.5   Strong Degeneracy

The Einstein-Bose and Fermi-Dirac distributions both approach the Maxwell-Boltzmann limit for $\zeta \cong 0$, but at the other extreme, where there is strong degeneracy, the two cases are quite different. Consider the F.D. law first. We have by (2.19)

$$n(\epsilon) = f(\epsilon)g(\epsilon) \tag{5.1}$$

where $g(\epsilon)$ is the degeneracy factor, and

$$f(\epsilon) = \frac{1}{\zeta^{-1}e^{\epsilon/kT} + 1} \tag{5.1a}$$

$$\zeta = e^{\epsilon_F/kT} \tag{5.1b}$$

where ϵ_F, the Fermi energy, is the Gibbs energy per particle. At $\epsilon = \epsilon_F$,

$$f(\epsilon) = \frac{1}{2}.$$

At $\epsilon = 0$, and large ζ,

$$f(\epsilon) \cong 1.$$

For large ϵ, the tail is Maxwellian. As the temperature is lowered $f(\epsilon)$ approaches a step function. The limiting form at $T = 0$ is illustrated in Figure 8.2.

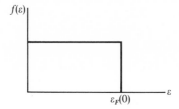

FIGURE 8.2
Fermi distribution at absolute zero
(complete degeneracy).

We now find by (2.23a) and Figure 8.2

$$F_1(\zeta) = 4\pi \int_0^\infty \frac{y^2 \, dy}{\zeta^{-1} e^{y^2} + 1}$$

$$\cong \frac{4\pi}{3} \bar{y}^3 \tag{5.2}$$

and by (2.22c) and Figure 8.2

$$F_2(\zeta) = \frac{8\pi}{3} \int_0^\infty \frac{y^4 \, dy}{\zeta^{-1} e^{y^2} + 1}$$

$$\cong \frac{8\pi}{3} \frac{\bar{y}^5}{5} \tag{5.3}$$

where $\bar{y}$ is the cut-off of the step function. Therefore by (4.3)

$$\frac{PV}{RT} \cong \frac{2}{5} \bar{y}^2. \tag{5.4}$$

But by (5.2)

$$\theta = F_1(\zeta) \cong 4\pi \frac{\bar{y}^3}{3}. \tag{5.5}$$

Hence

$$\frac{PV}{RT} = \frac{2}{5} \left(\frac{3\theta}{4\pi} \right)^{2/3} \sim \left(\frac{1}{VT^{3/2}} \right)^{2/3} \sim \frac{1}{T} V^{-2/3}$$

or

$$PV^{5/3} = \text{constant.} \tag{5.6}$$

The completely degenerate Fermi gas obeys the adiabatic law with $\gamma = 5/3$.

8.6 Examples of Degenerate Fermi-Dirac Systems

Electrons in metals and nucleons in nuclei are examples of F.D. gases. The properties of metals are decisively altered by the F.D. statistics. In particular the specific heat of the electron gas is much less than its classical value, $3/2\,R$

per mol. The small value of the specific heat may be calculated directly from our expression for A [equation (2.22)] and understood simply as follows. Electrons are, roughly speaking, unable to contribute to the specific heat unless they lie within the distance kT of the Fermi surface because of the exclusion principle. Hence the fraction of electrons contributing is kT/ϵ_F, and the resulting specific heat is

$$C_V \cong \frac{kT}{\epsilon_F} \cdot \frac{3}{2} R = \frac{3}{2} R \cdot \left(\frac{T}{T_F}\right) \tag{6.1}$$

where

$$\epsilon_F = kT_F. \tag{6.2}$$

In addition to being very small, C_V is directly proportional to T. The total specific heat of a metallic crystal is hence

$$C_V = a\left(\frac{T}{T_F}\right) + b\left(\frac{T}{\theta}\right)^3 \tag{6.3}$$

where T_F and θ are Fermi and Debye temperatures respectively. The magnitude and behavior of the paramagnetic susceptibility may be similarly understood. According to the classical Curie law,

$$\chi = \frac{1}{3} \frac{N\mu^2}{kT}. \tag{6.4}$$

But this is also diminished by the same factor T/T_F; hence the correct χ for a metal, as first noted by Pauli, is

$$\chi_{\text{Pauli}} \cong \frac{T}{T_F} \chi_{\text{classical}}$$

$$\cong \frac{1}{3} \frac{N\mu^2}{3T_F} \tag{6.5}$$

which is again small and now temperature independent.

These formulas for the specific heat and magnetic susceptibility are valid up to high temperatures, since kT ("room temperature") $\sim .025$ volt, while Fermi energies for metals are of the order of one volt.

It may be remarked again that the weak residual interaction responsible for superconductivity completely changes these results; the specific heat becomes exponential instead of linear, and instead of weak paramagnetism one observes perfect diamagnetism.

WHITE-DWARF STARS

The parameter θ is large for electrons in metals because m is small. The degeneracy parameter can also become large even at high temperatures if the density is high. This situation is realized in white dwarf stars where the density becomes of the order of 10^6, and would also be realized in neutron stars.

8.7   The Degenerate Nonrelativistic Einstein-Bose System (The Superfluid)

In the F.D. case the degeneracy parameter may become arbitrarily large, but in the E.B. case ζ can never exceed unity. For now

$$\langle n_s \rangle = g_s [\zeta^{-1} \exp (\epsilon_s/kT) - 1]^{-1} \qquad (7.1)$$

Since the $\langle n_s \rangle$ are necessarily positive,

$$\zeta^{-1} \exp (\epsilon_s/kT) - 1 \geqq 0 \qquad (7.2)$$

or

$$\zeta \leqq \exp (\epsilon_s/kT). \qquad (7.2a)$$

In particular this condition must be satisfied for the lowest level, $\epsilon_0 = 0$. Hence

$$\zeta \leqq 1 \qquad (7.3)$$

or

$$G \leqq 0. \qquad (7.3a)$$

In the E.B. case extreme degeneracy is therefore realized when $\zeta = 1$ or $G = 0$.

Consider a value of ζ near unity:

$$\zeta = 1 - \epsilon. \qquad (7.4)$$

Then

$$\langle n_0 \rangle = \frac{g_0}{\zeta^{-1} - 1} = \frac{g_0}{\epsilon}. \qquad (7.5)$$

Hence as ζ approaches unity all of the particles tend to fall into the lowest state. This behavior is called the E.B. condensation; it is a condensation in momentum space but not in configuration space. An estimate of the critical value of θ for which $\zeta = 1$ may be made from the equation (2.23):

$$\theta = 4\pi \int_0^\infty \frac{y^2 \, dy}{e^{y^2} - 1} \qquad (7.6)$$

or

$$\frac{\rho h^3}{(2\pi mkT_0)^{3/2}} = 2.6. \qquad (7.6a)$$

Putting in the density of liquid helium, one finds the critical temperature to be $T_0 \cong 3°$. The λ-transition, at 2.2°, is indeed an E.B. condensation. That the λ-transition must be understood in this way is most clearly illustrated by the fact that it is absent in He³; since He³ and He⁴, being chemically indistinguishable, differ only in their statistics, no other explanation is possible.

8.8   Specific Heat and Pressure of the Degenerate Nonrelativistic Einstein-Bose System

As the temperature is lowered, more particles condense into the ground state. We may estimate the number of particles (n^*) which have *not* condensed as follows:

$$n^* = \sum_{s>0} g_s [\zeta^{-1} \exp(\epsilon_s/kT) - 1]^{-1}$$

$$\cong \sum_{s>0} g_s [\exp(\epsilon_s/kT) - 1]^{-1}. \tag{8.1}$$

Again we may estimate the right side by an integral and find by (2.23)

$$\theta(n^*, T) = F_1(1). \tag{8.2}$$

We define the critical temperature T_0 by the conditions that $\zeta = 1$ and that $n^* = n$. Then

$$\theta(n, T_0) = F_1(1) \tag{8.3}$$

and therefore by (2.22b),

$$\frac{n^*}{n} = \left(\frac{T}{T_0}\right)^{3/2}. \tag{8.4}$$

The preceding equations are supposed to hold when $T/T_0 \leq 1$. At 0° all particles have condensed; at T_0 no particles have condensed. According to (2.22) we find at $\zeta = 1$

$$\frac{A}{RT} = \ln \zeta + \frac{1}{\theta} F_2(\zeta)$$

$$= \frac{1}{\theta} F_2(1). \tag{8.5}$$

Therefore

$$A \sim T^{5/2} V \tag{8.6}$$

and the two equations of state determined by (8.6) then imply the following:

$$C_V \sim T^{3/2} \tag{8.7}$$

$$P \sim T^{5/2}. \tag{8.8}$$

8.9 The Degenerate Relativistic Einstein-Bose System (The Photon Gas)

The photon gas is another example of extreme degeneracy:

$$\zeta = 1 \tag{9.1}$$

$$G = 0. \tag{9.1a}$$

Compare with equation (5.9) of Chapter 7. However this system is completely relativistic as well. We have

$$n = \sum_s g_s [\exp(\epsilon_s/kT) - 1]^{-1}. \tag{9.2}$$

Going over from the sum to the integral, one finds

$$n = \frac{8\pi V}{h^3} \int_0^\infty \frac{p^2 \, dp}{e^{pc/kT} - 1} \tag{9.3}$$

where the relativistic connection between energy and momentum has now been used. Then

$$n = \frac{8\pi V}{h^3} \int_0^\infty \frac{y^2 \, dy}{e^y - 1} \left(\frac{kT}{c}\right)^3 \tag{9.4}$$

where

$$\frac{pc}{kT} = y. \tag{9.5}$$

Therefore in this case the number of particles (photons) varies as T^3. By substituting (9.5) in (2.21) one obtains the corresponding formula

$$A \sim VT^4$$

and the result is that the energy content varies as T^4. (See equation (5.7), Chapter 7.)

The temperature dependence of the specific heat is different for the photon gas and the degenerate molecular gas, discussed in the preceding section (even though both obey Bose statistics with $\zeta = 1$) because the energy-momentum relation is relativistic in the photon case and it is nonrelativistic in the molecular case.

Formally these two cases differ in the relation between energy and momentum of the boson: the relativistic relation is linear and the nonrelativistic relation is quadratic. Similarly the relation is linear for phonons and therefore the Debye specific heat $\sim T^3$; while it is quadratic for spin waves [3] and the corresponding specific heat $\sim T^{3/2}$.

Helium II obeys the T^3 (not the $T^{3/2}$) law in spite of the result of (8.8); just as in superconductors this change in the specific heat law is caused by the weak interaction responsible for the superstate. The T^3 law follows from the fact that near $p = 0$, the excitations (phonons) in He II satisfy the relation

$$\epsilon = cp.$$

Therefore one has the "massless" or Debye type of behavior.

Notes and References

1. Section 8.2 is based on E. Schrodinger, *Statistical Thermodynamics* (Cambridge Univ. Press, Cambridge, 1948). The evaluation of the contour integral (2.10) is further discussed in Chapter 6 of this reference. This is the method of Darwin and Fowler.

2. This rule is equivalent to the familiar classical expression $V\,d\mathbf{k}$, which is the number of plane waves that fit into a cavity of volume V and lie in the interval $d\mathbf{k}$. The origin of this degeneracy is the equivalence of different directions or the isotropy of space.

   The general problem of determining the degeneracy function or density of states [g_n or $\rho(E)$ in (19.8) or (19.9) of Chapter 6] is of course that of determining the distribution of the eigenvalues of the Schrodinger equation for the complete system. In the elementary examples of noninteracting particles this problem is trivial since one merely has to count the number of plane waves that fit into a cavity. This is the same question one has to answer in obtaining the Rayleigh-Jeans formula.

3. These are collective excitations in a system of spins. The theory of such excitations is related to the corresponding theory of uncoupled spins similar to the way the Debye description of a lattice is related to the Einstein model. See, for example, G. Wannier, *Statistical Physics*, Chapter 15 (Wiley, New York, 1966).

9

The Grand Partition Function
and Second Quantization

9.1 The Grand Canonical Ensemble

The grand canonical ensemble provides the appropriate statistical representation of an open thermodynamic system, i.e., one which may exchange mass,
as well as energy and momentum, with its surroundings. The analysis may
again be based on the H-theorem, and thermodynamic equilibrium may be
correlated with an ensemble which minimizes $\bar{\bar{H}}$ under conditions of "essential isolation". In other words it is supposed that some arbitrary nonequilibrium ensemble evolves into an equilibrium ensemble by exchanging
energy and mass with its surroundings in such a way that some members of
the ensemble gain, and others lose, energy and mass without changing the
mean energy and mass for the entire ensemble. Therefore one seeks a minimum in $\bar{\bar{H}}$ subject to constraints on the total energy and mass:

$$\delta\bar{\bar{H}} = 0 \qquad \text{or} \qquad \sum_\alpha \delta P_\alpha \ln P_\alpha = 0$$

$$\delta\bar{E} = 0 \qquad \text{or} \qquad \sum_\alpha \delta P_\alpha E_\alpha = 0$$

$$\delta\bar{\bar{N}} = 0 \qquad \text{or} \qquad \sum_\alpha \delta P_\alpha N_\alpha = 0$$

$$\delta\left(\sum_\alpha P_\alpha\right) = 0 \tag{1.1}$$

where N denotes the number of mols. The mathematical problem is just the same as the one that arises in connection with the canonical ensemble, except for the additional constraint, which of course leads to an additional Lagrangian multiplier. The result is

$$P_\alpha = e^{(\Omega - E_\alpha)/\theta} e^{-cN_\alpha/\theta} \tag{1.2}$$

where c is the new multiplier. In order to interpret these constants $(\Omega, \theta, \text{and } c)$, one displaces the equilibrium by altering the external constraints, just as in the analysis of the canonical ensemble. One obtains

$$\Delta \bar{\bar{H}} = \sum_\alpha \Delta P_\alpha \ln P_\alpha$$

$$= \sum_\alpha \Delta P_\alpha \left(\frac{\Omega - E_\alpha - cN_\alpha}{\theta} \right)$$

$$-\theta \Delta \bar{\bar{H}} = \sum_\alpha \Delta P_\alpha (E_\alpha + cN_\alpha)$$

$$= \Delta \left(\sum_\alpha P_\alpha E_\alpha \right) - \sum_\alpha P_\alpha \left(\sum_s \frac{\partial E_\alpha}{\partial a_s} \Delta a_s \right) + c \sum_\alpha \Delta P_\alpha N_\alpha$$

where Δa_s are the changes in the constraints. Repeating the same argument as for the canonical ensemble, one finds the second law in the form:

$$-\theta \Delta \bar{\bar{H}} = \Delta U + \Delta W + c \sum_\alpha (\Delta P_\alpha) N_\alpha \tag{1.3}$$

where ΔU is the change in internal energy and ΔW is the work done by the system when the equilibrium is shifted. The third term on the right is

$$c \sum_\alpha \Delta P_\alpha N_\alpha = c \Delta \bar{\bar{N}} = c \Delta N$$

where N means $\bar{\bar{N}}$, and therefore (1.3) becomes

$$T \Delta S = \Delta U + \Delta W + c \Delta N \tag{1.4}$$

or

$$\Delta U = T \Delta S - P \Delta V - c \Delta N. \tag{1.4a}$$

Similarly,

$$\Delta G = -S \Delta T + V \Delta P - c \Delta N \tag{1.4b}$$

$$\Delta A = -S \Delta T - P \Delta V - c \Delta N. \tag{1.4c}$$

Hence

$$\left(\frac{\partial U}{\partial S} \right)_{NV} = T, \qquad \left(\frac{\partial U}{\partial V} \right)_{NS} = -P, \qquad \left(\frac{\partial U}{\partial N} \right)_{SV} = -c \tag{1.5}$$

and

$$c = -\left(\frac{\partial G}{\partial N}\right)_{TP} = -\left(\frac{\partial A}{\partial N}\right)_{TV}. \tag{1.6}$$

Therefore we shall write [1]

$$P_\alpha = e^{(\Omega - E_\alpha)/kT} e^{\mu N_\alpha/kT} \tag{1.7}$$

where

$$\mu = \left(\frac{\partial G}{\partial N}\right)_{TP} = \left(\frac{\partial A}{\partial N}\right)_{TV} = \left(\frac{\partial U}{\partial N}\right)_{SV}. \tag{1.8}$$

By the normalization condition,

$$\Omega = -kT \ln Q \tag{1.9}$$

where Q is the grand partition function, namely,

$$Q \equiv \sum_\alpha e^{-E_\alpha/kT} e^{\mu N_\alpha/kT}. \tag{1.9a}$$

and the α sum is over all states. Again

$$\bar{\bar{H}} = \sum_\alpha P_\alpha \ln P_\alpha$$

By (1.7)

$$-TS = \sum_\alpha P_\alpha(\Omega - E_\alpha + \mu N_\alpha)$$

or

$$\Omega = U - TS - \mu N = A - \mu N. \tag{1.10}$$

Hence by (1.9)

$$A = \mu N - kT \ln Q. \tag{1.10a}$$

It follows from (1.9a) that the grand partition function may also be expressed as follows:

$$Q = \text{Tr } e^{-H/kT} e^{\mu N/kT} \tag{1.11}$$

where Tr means trace of the quantum operators formed from the Hamiltonian H and the number operator N. It has been assumed that H and N commute and may therefore be diagonalized simultaneously.

Another important representation that follows from (1.9a) is the expansion

$$Q = \sum_{N_\alpha} Q_{N_\alpha} \zeta^{N_\alpha} \tag{1.12}$$

where Q_{N_α} is the canonical partition function of a system with N_α particles and where

$$\zeta = e^{\mu/kT}. \tag{1.13}$$

It is also interesting to express P_α directly in terms of the thermodynamic variables. From (1.7) and (1.10)

$$P_\alpha = e^{(A-E_\alpha)/kT} \, e^{\frac{\mu}{kT}(N_\alpha-N)}$$

$$= e^{(A-U)/kT} \, e^{(U-E_\alpha)/kT} \, e^{\frac{\mu}{kT}(N_\alpha-N)}$$

or

$$P_\alpha = e^{-S/k} \, e^{-\Delta E_\alpha/kT} \, e^{\frac{\mu}{kT}\Delta N_\alpha} \tag{1.14}$$

where

$$\Delta E_\alpha = E_\alpha - U$$
$$\Delta N_\alpha = N_\alpha - N.$$

Equation (1.14) permits one to compare the three kinds of canonical ensemble. In the *grand canonical* ensemble, the dispersions in E_α and N_α are measured by kT and kT/μ. The *canonical* ensemble shows only a spread in energy, and the *microcanonical* distribution has a sharp energy as well as a sharp particle number. In this last case

$$P_\alpha = e^{-S/k} \tag{1.15}$$

and the normalization condition reads

$$ge^{-S/k} = 1 \tag{1.16}$$

which is the same as Boltzmann's relation:

$$S = k \ln g \tag{1.16a}$$

where g is the total number of states.

9.2   Expectation Values and Fluctuations

In this section we shall make the change of notation $N_\alpha \longrightarrow N$ and $N \longrightarrow \langle N \rangle$. Returning to (1.12) we find

$$\zeta \frac{\partial Q}{\partial \zeta} = \sum_N N Q_N \zeta^N$$

and

$$\langle N \rangle = \frac{\sum\limits_N N Q_N \varsigma^N}{\sum\limits_N Q_N \varsigma^N} = \frac{\varsigma}{Q} \frac{\partial Q}{\partial \varsigma}.$$

Therefore the expectation value of N is

$$\langle N \rangle = \frac{\partial \ln Q}{\partial \ln \varsigma}. \tag{2.1}$$

By repeating this procedure we may find successive moments. For example,

$$\varsigma \frac{\partial}{\partial \varsigma} \cdot \varsigma \frac{\partial}{\partial \varsigma} Q = \sum\limits_N N^2 Q_N \varsigma^N$$

or

$$\langle N^2 \rangle = \frac{\sum\limits_N N^2 Q_N \varsigma^N}{\sum\limits_N Q_N \varsigma^N}$$

$$= \frac{1}{Q} \frac{\partial^2 Q}{\partial (\ln \varsigma)^2}.$$

Let us next calculate

$$\langle (\Delta N)^2 \rangle = \langle (N - \langle N \rangle)^2 \rangle$$
$$= \langle N^2 \rangle - 2\langle N \rangle \langle N \rangle + \langle N \rangle^2$$
$$= \langle N^2 \rangle - \langle N \rangle^2.$$

But

$$\frac{\partial^2 (\ln Q)}{\partial (\ln \varsigma)^2} = \frac{\partial}{\partial (\ln \varsigma)} \frac{1}{Q} \frac{\partial Q}{\partial (\ln \varsigma)}$$

$$= -\frac{1}{Q^2} \left(\frac{\partial Q}{\partial \ln \varsigma} \right)^2 + \frac{1}{Q} \frac{\partial^2 Q}{\partial (\ln \varsigma)^2}$$

$$= -\langle N \rangle^2 + \langle N^2 \rangle.$$

Therefore

$$\langle (\Delta N)^2 \rangle = \frac{\partial^2 \ln Q}{\partial (\ln \varsigma)^2}. \tag{2.2}$$

There are similar formulas for the expectation value and fluctuations in the energy. Let

$$\chi = e^{-1/kT}. \tag{2.3}$$

Since (χ, E) appear in the same way as (ζ, N) in the expansion of the grand partition function, according to (1.9a) and [3], we have

$$U = \langle E \rangle = \frac{\partial \ln Q}{\partial \ln \chi} \tag{2.4}$$

$$\langle (\Delta E)^2 \rangle = \frac{\partial^2 \ln Q}{\partial (\ln \chi)^2}. \tag{2.5}$$

It follows from (2.1)–(2.5) that

$$U = -k \frac{\partial \ln Q}{\partial (1/T)} = kT^2 \left(\frac{\partial \ln Q}{\partial T} \right)_{\zeta V} \tag{2.6}$$

$$\langle (\Delta E)^2 \rangle = \frac{\partial U}{\partial (\ln \chi)} = kT^2 \left(\frac{\partial U}{\partial T} \right)_{\zeta V} = kT^2 C_V \tag{2.7}$$

$$\langle N \rangle = \left(\frac{\partial \ln Q}{\partial \ln \zeta} \right)_{\chi V} = kT \left(\frac{\partial \ln Q}{\partial \mu} \right)_{TV} \tag{2.8}$$

$$\langle (\Delta N)^2 \rangle = \left(\frac{\partial \langle N \rangle}{\partial \ln \zeta} \right)_{\chi V} = kT \left(\frac{\partial \langle N \rangle}{\partial \mu} \right)_{TV}. \tag{2.9}$$

9.3   Equations of State from Grand Partition Function

The pressure, energy density, and mass density may all be obtained from the free energy (1.10a):

$$A(\mu, T, V) = \mu N - kT \ln Q \tag{3.1}$$

where we have returned to the notation of Section 9.1. Notice first

$$\left(\frac{\partial A}{\partial \mu} \right)_{TV} = \mu \left(\frac{\partial N}{\partial \mu} \right)_{TV} + N - kT \left(\frac{\partial}{\partial \mu} \ln Q \right)_{TV}$$

$$= \mu \frac{\langle (\Delta N)^2 \rangle}{kT} + N - N = \mu \frac{\langle (\Delta N)^2 \rangle}{kT}. \tag{3.2a}$$

Therefore in the approximation of a sharp distribution in particle number,

$$\left(\frac{\partial A}{\partial \mu} \right)_{TV} = 0. \tag{3.2b}$$

This is the side condition (2.13a) of Chapter 8. On the other hand the Gibbs function of a homogeneous system is

$$G(P, T) = \mu(P, T)N \tag{3.3}$$

and therefore by (3.1)

$$A = G - kT \ln Q$$

or

$$PV = kT \ln Q \tag{3.4}$$

where V is the volume of the system. In addition we have, by (2.6) and (2.8),

$$N = kT \frac{\partial \ln Q}{\partial \mu} \tag{3.5}$$

$$U = kT^2 \frac{\partial \ln Q}{\partial T}. \tag{3.6}$$

These equations (3.4), (3.5), (3.6) may also be rewritten as follows by (1.9):

$$P = -\frac{\Omega}{V} \tag{3.7}$$

$$N = -\frac{\partial \Omega}{\partial \mu} \tag{3.8}$$

$$U = -T^2 \frac{\partial}{\partial T} \left(\frac{\Omega}{T} \right). \tag{3.9}$$

All of these equations for P, N, and U depend parametrically on ζ. The thermal equation of state is given by the pair of equations for $P(\zeta)$ and $N(\zeta)$, while the caloric equation is similarly obtained by eliminating ζ from $U(\zeta)$ and $N(\zeta)$.

9.4 Thermodynamic Stability of Open Systems

The thermal and caloric equations of state satisfy the usual compatibility relations because of the way they are derived from Ω. They must also satisfy the conditions of thermodynamic stability, such as [2]

$$\left(\frac{\partial U}{\partial T} \right)_{NV} > 0 \tag{4.1}$$

$$\left(\frac{\partial N}{\partial \mu} \right)_{UV} > 0 \tag{4.2}$$

$$\left(\frac{\partial P}{\partial \rho} \right)_{T} > 0. \tag{4.3}$$

These conditions are imposed by the requirements of stability against energy, mass, and momentum transport respectively. The first two relations are nearly established by (2.7) and (2.9):

$$\left(\frac{\partial U}{\partial T}\right)_{\mu V} = \frac{1}{kT^2}\langle(\Delta E)^2\rangle > 0 \tag{4.1a}$$

$$\left(\frac{\partial N}{\partial \mu}\right)_{TV} = \frac{1}{kT}\langle(\Delta N)^2\rangle > 0 \tag{4.2a}$$

since these averages are positive.

However conditions (4.1) and (4.2) require that N or U be held fixed. To go from (4.1a) and (4.2a) to (4.1) and (4.2) consider

$$\Delta U = \frac{\partial U}{\partial(\ln \chi)}\,d(\ln \chi) + \frac{\partial U}{\partial(\ln \zeta)}\,d(\ln \zeta)$$

$$\Delta N = \frac{\partial N}{\partial(\ln \chi)}\,d(\ln \chi) + \frac{\partial N}{\partial(\ln \zeta)}\,d(\ln \zeta)$$

from which we obtain

$$\left(\frac{\partial U}{\partial \ln \chi}\right)_{NV} = \Delta \bigg/ \left(\frac{\partial N}{\partial \ln \zeta}\right)_{TV}$$

$$\left(\frac{\partial N}{\partial \ln \zeta}\right)_{UV} = \Delta \bigg/ \left(\frac{\partial U}{\partial \ln \chi}\right)_{\mu V}$$

where

$$\Delta = \frac{\partial U}{\partial \ln \chi}\frac{\partial N}{\partial \ln \zeta} - \frac{\partial U}{\partial \ln \zeta}\frac{\partial N}{\partial \ln \chi}$$

$$= \left[\frac{\partial^2 \ln Q}{\partial(\ln \chi)^2}\right]\left[\frac{\partial^2 \ln Q}{\partial(\ln \zeta)^2}\right] - \left[\frac{\partial^2 \ln Q}{\partial(\ln \zeta)(\ln \chi)}\right]^2. \tag{4.4}$$

The discriminant Δ, which is negative for saddle points and positive for maxima and minima, is here positive since $\ln Q$ is a minimum with respect to both χ and ζ [3]. Therefore (4.1) and (4.2) follow from (4.1a) and (4.2a).

To obtain the third condition, let us introduce the particle density

$$\rho = \frac{N}{V}. \tag{4.5}$$

Then the relation between P and ρ is obtained by elimination of ζ from the following pair of equations:

$$P = \frac{kT}{V}\ln Q(\zeta) \tag{4.6a}$$

$$\rho = \frac{N(\zeta)}{V}. \tag{4.6b}$$

Therefore

$$\left(\frac{\partial P}{\partial \rho}\right)_T = \left(\frac{\partial P}{\partial \zeta}\right)_T \left(\frac{\partial \zeta}{\partial \rho}\right)_T.$$

By (4.6a) and (2.8)

$$\left(\frac{\partial P}{\partial \zeta}\right)_T = \frac{kT}{V} \frac{\partial}{\partial \zeta} \ln Q(\zeta)$$

$$= \frac{kT}{V} \frac{N}{\zeta}.$$

By (4.6b) and (2.9)

$$\left(\frac{\partial \rho}{\partial \zeta}\right)_T = \frac{1}{V} \left(\frac{\partial N}{\partial \zeta}\right)_T$$

$$= \frac{1}{\zeta} \frac{1}{V} \frac{\partial N}{\partial \ln \zeta} = \frac{1}{\zeta} \frac{1}{V} \langle (\Delta N)^2 \rangle$$

and therefore

$$\left(\frac{\partial P}{\partial \rho}\right)_T = kT \frac{N}{\langle (\Delta N)^2 \rangle} > 0. \tag{4.3a}$$

These relations (4.1a), (4.2a), and (4.3a) are required for thermodynamic stability. They may also be used to estimate the magnitude of the fluctuations in N and E in terms of thermodynamically observable quantities. In particular we remark that near the critical point the coefficient $(\partial P/\partial \rho)_T$ becomes very small and therefore by (4.3a) the fluctuations become very large. These large fluctuations give rise to critical opalescence, for example.

9.5   Grand Partition Function of Ideal Gas

The formalism of the grand canonical ensemble is not only of more general applicability, but even in cases where the number of particles may be regarded as fixed, it is more powerful than the canonical formalism. One may illustrate this point by calculating the properties of an ideal quantum mechanical gas.

It is necessary to calculate

$$Q = \sum_\alpha e^{-E_\alpha/kT} e^{\mu N_\alpha/kT} \tag{5.1}$$

where

$$E_\alpha = \sum_s n_s \epsilon_s \tag{5.1a}$$

$$N_\alpha = \sum_s n_s. \tag{5.1b}$$

Here ϵ_s is the energy of a single particle state, and n_s is the number of particles in that state. We find

$$Q = \sum_{n_1, n_2, \ldots} e^{-(n_1\epsilon_1 + n_2\epsilon_2 + \cdots)/kT} \, e^{(n_1 + n_2 + \cdots)\mu/kT}$$

$$= \sum_{n_1, n_2, \ldots} x_1^{n_1} \, x_2^{n_2} \cdots \zeta^{n_1} \zeta^{n_2} \cdots$$

$$= \sum_{n_1, n_2, \ldots} (x_1\zeta)^{n_1} \, (x_2\zeta)^{n_2} \cdots$$

$$= Q_\infty(\zeta x_1, \zeta x_2, \cdots). \tag{5.2}$$

Therefore by (3.1) and (3.2b)

$$A = \mu N - kT \ln Q_\infty(\zeta x_1, \zeta x_2, \ldots) \tag{5.3}$$

$$\frac{\partial A}{\partial \mu} = 0.$$

This result agrees with the expression obtained earlier from the canonical partition function, namely,

$$Q_\infty(\zeta x_1, \zeta x_2, \ldots) = \Pi(1 \pm \zeta x_s)^{\pm 1}.$$

[See Chapter 8, equation (2.8).] In addition one finds
(a) *Equations of State*:

$$PV = kT \ln Q_\infty(\zeta x_1 \cdots) \tag{5.4}$$

$$N = \frac{\partial \ln Q_\infty(\zeta x_1 \cdots)}{\partial \ln \zeta} \tag{5.5}$$

$$U = \frac{\partial \ln Q_\infty(\zeta x_1 \cdots)}{\partial \ln \chi}. \tag{5.6}$$

(b) *Fluctuations*:

Let us limit our attention to particles of one energy. Then by (2.19) of Chapter 8,

$$n_s = \frac{g_s}{\zeta^{-1} x_s \pm 1} \tag{5.7}$$

$$\langle (\Delta n_s)^2 \rangle = \frac{\partial n_s}{\partial \ln \zeta}$$

$$= \frac{\partial n_s}{\partial \zeta^{-1}} \frac{\partial \zeta^{-1}}{\partial \ln \zeta}$$

$$= \frac{- g_s x_s}{(\zeta^{-1} x_s \pm 1)^2} \frac{\partial \zeta^{-1}}{\partial \ln \zeta}$$

$$= \left(\frac{n_s}{g_s}\right)^2 (g_s \, \zeta^{-1} x_s). \tag{5.7a}$$

Therefore the fractional fluctuation is

$$\frac{\langle(\Delta n_s)^2\rangle}{n_s^2} = \frac{1}{g_s}\,(\zeta^{-1}\,x_s)$$

$$= \frac{1}{n_s}\left(1 \mp \frac{n_s}{g_s}\right). \tag{5.8}$$

In the classical limit,

$$\frac{\langle(\Delta n_s)^2\rangle}{n_s^2} = \frac{1}{n_s}. \tag{5.9}$$

Fluctuations are therefore suppressed by the F.D. statistics and enhanced by the E.B. statistics.

In the E.B. condensation, n_0 becomes very large and hence

$$\frac{\langle(\Delta n_0)^2\rangle}{n_0^2} \simeq \frac{1}{g_0} \tag{5.10}$$

independent of n_0. In this case the fluctuation in n_0 is increased by the very large factor n_0/g_0 over the classical value, $1/n_0$, in agreement with the general rule that fluctuations become large at a phase transition.

9.6   Operator Formalism for Grand Partition Function (Second Quantization)

Ideal systems are of course mathematically trivial, and the real usefulness of the grand partition function is tested only by systems of interacting particles. In these problems the grand partition function is most fully exploited by the use of an operator formalism (second quantization) which will now be briefly explained.

To discuss an ideal gas it was found convenient, even in Chapter 8, to choose as independent variables the occupation numbers of the single particle states—rather than the coordinates of the particles. In an ideal gas these occupation numbers are constants, but if there is interaction, there is also scattering, and the number of particles in a given single particle state becomes time dependent. The occupation numbers then become dynamical variables.

The interaction may be described most directly by the matrix elements which describe the scattering of particles from given states of initial momentum to other states of final momentum. At the same time it is very useful to introduce certain operators (absorption operators) which remove particles from the initial state and other operators (emission operators) which place them in the final state, and to express H and N in terms of these more primitive operators. The grand partition function then becomes by (1.11)

$$Q = \text{Tr } e^{-H/kT} e^{\mu N/kT} \tag{6.1}$$

where now both H and N are written in terms of the emission and absorption operators. The grand partition function when so expressed provides the usual starting point for the quantum mechanical many-body formalism.

EMISSION AND ABSORPTION OPERATORS

These new operators will be denoted by $\bar{a}$(emission) and a(absorption) where $\bar{a}$ is the Hermitian adjoint of a. They are defined by the following relations:

$$(a_k, \bar{a}_l)_\pm = \delta_{kl} \tag{6.2a}$$

$$(a_k, a_l)_\pm = (\bar{a}_k, \bar{a}_l)_\pm = 0 \tag{6.2b}$$

where the range of the indices (k, l) is so far not specified, and where

$$(a, b)_- = ab - ba \quad \text{(commutator)}$$

$$(a, b)_+ = ab + ba \quad \text{(anticommutator)}.$$

The operators satisfying the $(\pm)$-rules are designated Fermi-Dirac and Einstein-Bose, respectively, for reasons which will become clear.

Now consider the operator

$$N_k = \bar{a}_k a_k. \tag{6.3}$$

Then

$$(N_k, a_l)_- = (\bar{a}_k a_k, a_l)_-.$$

This commutator may be evaluated with the aid of the following identities:

$$(ab, c)_- = a(b, c)_- + (a, c)_- b$$

$$= a(b, c)_+ - (a, c)_+ b.$$

Then

$$(N_k, a_l)_- = \bar{a}_k (a_k, a_l)_- + (\bar{a}_k, a_l)_- a_k = - a_k \delta_{kl} \quad \text{E.B.}$$

$$= \bar{a}_k (a_k, a_l)_+ - (\bar{a}_k, a_l)_+ a_k = - a_k \delta_{kl}. \quad \text{F.D.}$$

Therefore in both cases

$$(N_k, a_l)_- = - a_k \delta_{kl}. \tag{6.4a}$$

Similarly, by taking the Hermitian adjoint of the preceding equation, one sees that

$$(N_k, \bar{a}_l) = \bar{a}_k \delta_{kl} \tag{6.4b}$$

for both E.B. and F.D. operators.

Next introduce the particular state $|0\rangle$, such that

$$a_k|0\rangle = 0|0\rangle \tag{6.5}$$

for all k. Then $|0\rangle$ will be called the vacuum state.

It follows from (6.5) that

$$N_k|0\rangle = 0|0\rangle \tag{6.6}$$

and therefore that $|0\rangle$ is an eigenstate of all the N_k.

Next let $|n_k\rangle$ be an arbitrary eigenstate of N_k. Then

$$N_k|n_k\rangle = n_k|n_k\rangle \tag{6.7}$$

where n_k is the corresponding eigenvalue. It then follows by (6.4b) that

$$(N_k\bar{a}_k - \bar{a}_kN_k)\,|n_k\rangle = \bar{a}_k|n_k\rangle$$

and by (6.7) that

$$N_k(\bar{a}_k|n_k\rangle) = (n_k + 1)(\bar{a}_k|n_k\rangle). \tag{6.8a}$$

Similarly,

$$N_k(a_k|n_k\rangle) = (n_k - 1)(a_k|n_k\rangle). \tag{6.8b}$$

According to (6.8a) and (6.8b), one concludes that $\bar{a}_k|n_k\rangle$ and $a_k|n_k\rangle$ are eigenstates belonging to the eigenvalues $n_k + 1$ and $n_k - 1$, respectively, if $|n_k\rangle$ is known to be an eigenstate belonging to the eigenvalue n_k. Therefore starting from the postulated eigenstate $|0\rangle$, one may construct other eigenstates of N_k by successively applying the operator $\bar{a}_k$ unless a vanishing eigenvector is obtained at some stage. In the E.B. case a zero eigenvector never appears, but in the F.D. case it does, as one can see by equation (6.2b):

$$\bar{a}_k\bar{a}_k + \bar{a}_k\bar{a}_k = 0$$

or

$$\bar{a}_k\bar{a}_k = 0. \tag{6.9}$$

Therefore in the F.D. case the only eigenvectors obtained in this way are $|0\rangle$ and $\bar{a}_k|0\rangle$ with $n_k = 0,1$. On the other hand, in the E.B. case there are an infinite number of eigenvectors, and their eigenvalues are $n_k = 0, 1, 2, \ldots$. It may be shown that all eigenvectors are obtained by this procedure.

Finally it follows from (6.4) that

$$(N_k, a_l) = (N_k, \bar{a}_l) = 0 \qquad l \neq k. \tag{6.10}$$

Therefore $a_l|n_k\rangle$ and $\bar{a}_l|n_k\rangle$ are also eigenstates of N_k belonging to the same eigenvalue (n_k).

Since they are integers, we may interpret the n_k as the numbers of particles in certain single particle states, and we may therefore choose the $N_k = \bar{a}_k a_k$ as the corresponding number operators. Since a_k and $\bar{a}_k$ decrease and increase n_k by one unit, respectively, they may be regarded as absorption and emission operators for the kth state. Since $n_k = 0, 1$ in the anticommuting case and $n_k = 0, 1, 2, \ldots$ in the commuting case, the F.D. statistics correspond to anticommutators and the E.B. statistics to commutators.

In the most important case the states k are the different possible one-particle momentum states. An eigenstate of an ideal quantum gas may then be written in the following way:

$$|n_1, n_2, \ldots\rangle = \bar{a}(k_1)^{n_1} \bar{a}(k_2)^{n_2} \ldots |0\rangle \tag{6.11}$$

where $(n_1, n_2, \ldots)$ are the populations of the one-particle momentum states $(k_1, k_2, \ldots)$.

Since these states form a complete set, an arbitrary state $(|\alpha\rangle)$ of a nonideal gas may be represented as a superposition of these states as follows:

$$|\alpha\rangle = \sum_{n_1, n_2, \ldots} |n_1, n_2, \ldots\rangle \langle n_1, n_2, \ldots |\alpha\rangle \tag{6.12}$$

where $|\langle n_1, n_2, \ldots |\alpha\rangle|^2$ is the probability of finding the particular distribution $(n_1, n_2, \ldots)$ in the arbitrary state $|\alpha\rangle$.

HAMILTONIAN AND NUMBER OPERATOR

One may now write

$$N = \sum_k \bar{a}_k a_k \tag{6.13}$$

$$H = \sum_k \epsilon_k \bar{a}_k a_k + U \tag{6.14}$$

where ϵ_k is the kinetic energy corresponding to the momentum k and U is the potential energy operator. It is often useful to express U as follows:

$$U = \frac{\lambda}{2} \sum_{k,l} (ij|u|kl)\, \bar{a}_i \bar{a}_j a_k a_l \tag{6.15}$$

where $\lambda(ij|u|kl)$ measures the matrix element for the scattering, $k + 1 \longrightarrow i + j$. Here λ is a parameter (which may be called the coupling constant).

DENSITY MATRIX AND EXPECTATION VALUES

The density matrix may be referred to this same basis which diagonalizes all the N_k. Then by (21.7) of Chapter 6

$$\langle n_1, \ldots |\rho| n'_1, \ldots \rangle = \sum_\alpha \langle n_1, \ldots |\alpha\rangle\, W_\alpha \langle \alpha| n'_1, \ldots \rangle \tag{6.16}$$

where the weights which determine the incoherent mixing of pure states are the temperature dependent W_α.

The average occupation numbers of the single particle states may be expressed in terms of ρ and formula (21.8) of Chapter 6 as follows:

$$\langle N_k \rangle = \text{Tr } \rho \, N_k$$
$$= \text{Tr } \rho \, \bar{a}_k a_k$$

or

$$\langle N_k \rangle = \text{Tr } a_k \, \rho \, \bar{a}_k. \tag{6.17}$$

One may then introduce the matrix ρ_1, such that

$$\langle k|\rho_1|l \rangle = \text{Tr } a_k \, \rho \, \bar{a}_l. \tag{6.18}$$

Then the diagonal values of this matrix are the average populations of the different one-particle states according to (6.17).

One may continue in this way to define other reduced density matrices, as follows:

$$\langle kl|\rho_2|ij \rangle = \text{Tr } a_k a_l \, \rho \, \bar{a}_i \bar{a}_j. \tag{6.19}$$

Finally note that when ρ describes a pure state, say $|\alpha\rangle$, then

$$\rho = |\alpha\rangle \langle \alpha|$$

and

$$\langle k|\rho_1|l \rangle = \text{Tr } a_k |\alpha\rangle \langle \alpha|\bar{a}_l$$
$$= \sum_\beta \langle \beta|a_k|\alpha\rangle \langle \alpha|\bar{a}_l|\beta\rangle$$
$$= \langle \alpha|\bar{a}_l a_k|\alpha\rangle. \tag{6.20}$$

Similarly,

$$\langle kl|\rho_2|ij \rangle = \langle \alpha|\bar{a}_i \bar{a}_j a_k a_l|\alpha\rangle. \tag{6.21}$$

FIELD OPERATOR

It is useful to introduce

$$\psi(\mathbf{x}) = \sum_k a_k \langle \mathbf{x}|k \rangle. \tag{6.22}$$

If $|\mathbf{k}\rangle$ is a one-particle momentum state,

$$\langle \mathbf{x}|\mathbf{k} \rangle = \frac{1}{V^{1/2}} e^{i\mathbf{k}\mathbf{x}}. \tag{6.22a}$$

In general $|\mathbf{k}\rangle$ is just some complete orthonormal set. Since the a_k are operators, $\psi(\mathbf{x})$ is an operator that also depends on position and is called a field operator.

We shall be interested in the following correlation coefficient, the expectation value of $\bar{\psi}(\mathbf{x}) \psi(\mathbf{y})$ for a macroscopic state described by the density matrix (ρ):

$$\langle \bar{\psi}(\mathbf{x}) \psi(\mathbf{y}) \rangle = \text{Tr } \rho \, \bar{\psi}(\mathbf{x}) \psi(\mathbf{y})$$

$$= \text{Tr } \psi(\mathbf{y}) \, \rho \, \bar{\psi}(\mathbf{x});$$

by (6.22),

$$= \sum_{k,l} \langle \mathbf{y}|\mathbf{l}\rangle \, (\text{Tr } a_l \, \rho \, \bar{a}_k) \, \langle \mathbf{k}|\mathbf{x}\rangle,$$

and by (6.18),

$$= \sum_{k,l} \langle \mathbf{y}|\mathbf{l}\rangle \, \langle \mathbf{l}|\rho_1|\mathbf{k}\rangle \, \langle \mathbf{k}|\mathbf{x}\rangle.$$

Therefore the two point correlation function is simply related to ρ_1 as follows:

$$\langle \bar{\psi}(\mathbf{x}) \psi(\mathbf{y}) \rangle = \langle \mathbf{y}|\rho_1|\mathbf{x}\rangle. \tag{6.23}$$

In a similar way, ρ_2 may be interpreted in terms of a four point correlation, and in general the reduced density matrices are expressed simply as correlation functions between field operators at different points.

The Hamiltonian and number operators may also be conveniently expressed in terms of $\psi(\mathbf{x})$.

The number operator is

$$\int \bar{\psi}(\mathbf{x}) \psi(\mathbf{x}) \, d\mathbf{x} = \sum_{k,l} \bar{a}(\mathbf{k}) \, a(\mathbf{l}) \int \langle \mathbf{k}|\mathbf{x}\rangle \, \langle \mathbf{x}|\mathbf{l}\rangle \, d\mathbf{x}$$

$$= \sum_{k} \bar{a}(\mathbf{k}) \, a(\mathbf{k}). \tag{6.24}$$

The kinetic energy contribution is

$$\int \bar{\psi}(\mathbf{x}) \, T(\mathbf{x}) \, \psi(\mathbf{x}) \, d\mathbf{x} = \sum_{k} \bar{a}(\mathbf{k}) \, a(\mathbf{k}) \, \epsilon(\mathbf{k}) \tag{6.25}$$

where

$$T(\mathbf{x}) = -\frac{\hbar^2}{2m} \, \nabla^2$$

and

$$\epsilon(\mathbf{k}) = \frac{\hbar^2 k^2}{2m}.$$

We may assume an interaction of the following form:

$$\frac{1}{2} \iint \bar{\psi}(x)\,\bar{\psi}(y)\,u(x-y)\,\psi(y)\,\psi(x)\,dx\,dy$$

$$= \frac{\lambda}{2} \sum_{k,l} \bar{a}(k)\,\bar{a}(l)\,a(p)\,a(q)\,\langle kl|u|pq\rangle \qquad (6.26)$$

where

$$\langle kl|u|pq\rangle = \iint dx\,dy\,\langle k|x\rangle\,\langle l|y\rangle\,u(x-y)\,\langle y|p\rangle\,\langle x|q\rangle. \qquad (6.26a)$$

Therefore the Hamiltonian may be expressed in either of the two equivalent forms:

$$H = \sum_k \bar{a}(k)\,a(k)\,\epsilon(k) + \frac{\lambda}{2}\sum_{k,l}\langle kl|u|pq\rangle\,\bar{a}(k)\,\bar{a}(l)\,a(p)\,a(q) \qquad (6.27a)$$

or

$$H = \int \bar{\psi}(x)\,T(x)\,\psi(x)\,dx + \frac{\lambda}{2}\iint dx\,dy\,\bar{\psi}(x)\,\bar{\psi}(y)\,u(x-y)\,\psi(y)\,\psi(x), \quad (6.27b)$$

while the number operator is

$$N = \int \bar{\psi}(x)\,\psi(x)\,dx. \qquad (6.28)$$

Notice finally that

$$\langle kl|u|pq\rangle = \frac{1}{V^2}\iint dx\,dy\,e^{i(py+qx-kx-ly)}\,u(x-y), \qquad (6.28a)$$

and after transformation to center of mass and relative coordinates:

$$x = X + \frac{r}{2}$$

$$y = X - \frac{r}{2}$$

$$\langle kl|u|pq\rangle = \frac{1}{V}\left[\int dr\,e^{i(q-k)r}\,u(r)\right]\delta(p+q-k-l). \qquad (6.29)$$

The δ-function of (6.29) permits a slight simplification of (6.27a):

$$H = \sum_k \bar{a}(k)\,a(k)\,\epsilon(k) + \frac{1}{2V}\sum_{k,l,n}\phi(n)\,\bar{a}(k)\,\bar{a}(l)\,a(l-n)\,a(k+n) \qquad (6.30)$$

where

$$\phi(n) = \lambda\int dr\,e^{inr}\,u(r). \qquad (6.30a)$$

The conservation of momentum is built into (6.30) because the interaction $u(\mathbf{x} - \mathbf{y})$ is translationally invariant. Here $\mathbf{n}$ is the momentum transfer $\mathbf{q} - \mathbf{k}$.

Notes and References

1. In this paragraph do not confuse capital rho with the pressure.
2. In Section 5.8, see equations (8.9), (8.11), and under (8.15) the relation $U_{MM} > 0$.
3. The discriminant Δ may be shown to be positive as follows. Start from the representation:

$$Q = \sum_{E_N N} \chi^{E_N} \zeta^N.$$

By straightforward calculation one finds

$$\frac{\partial}{\partial \ln \zeta} \frac{\partial}{\partial \ln x} \ln Q = -\frac{1}{Q^2} \frac{\partial Q}{\partial \ln \zeta} \frac{\partial Q}{\partial \ln x} + \frac{1}{Q} \frac{\partial^2 Q}{\partial (\ln \zeta) \, \partial (\ln x)}$$

$$= -\langle E \rangle \langle N \rangle + \langle EN \rangle$$

$$= \langle (\Delta E)(\Delta N) \rangle.$$

Therefore, by (2.2), (2.5), and (4.4) of Chapter IX,

$$\Delta = \langle (\Delta E)^2 \rangle \langle (\Delta N)^2 \rangle - \langle (\Delta E)(\Delta N) \rangle^2.$$

Let

$$\langle (\Delta E)^2 \rangle = \sum_i (\Delta E_i)^2 P_i = \sum_i u_i^2$$

$$\langle (\Delta N)^2 \rangle = \sum_i (\Delta N_i)^2 P_i = \sum_i v_i^2$$

where

$$u_i = (\Delta E_i) P_i^{1/2}$$
$$v_i = (\Delta N_i) P_i^{1/2}.$$

Then

$$\langle (\Delta E)(\Delta N) \rangle = \sum_i (\Delta E_i)(\Delta N_i) P_i$$

$$= \sum_i u_i v_i$$

and

$$\Delta = \left(\sum_i u_i^2 \right) \left(\sum_i v_i^2 \right) - \left(\sum_i u_i v_i \right)^2 \geq 0$$

by the Schwartz inequality.

10

Phase Transitions in Statistical Mechanics

10.1 General Remarks on the Problem of Phase Transitions

According to statistical mechanics, all thermodynamic properties may be deduced directly from knowledge of atomic structure with the aid of Schroedinger's equation and the partition function. Therefore all the distinct equations of state of the separate phases in which a given atomic structure may be realized should fall out of the general formalism, and of course all the phase transitions must be deducible as well. The main problem is to understand how, for example, the very different equations of state of the solid and gas arise from one and the same Hamiltonian; but again this problem is qualitatively similar to that of obtaining the properties of an atom and the corresponding unbound system from the same Hamiltonian.

To formulate the problem more precisely, consider a mass of hydrogen. Its spectrum will depend partly on external parameters such as the pressure. Suppose that the pressure is so fixed that the macroscopic system exists only in the solid and vapor forms. The problem is to determine the sublimation curve from the spectrum existing under this constraint.

The following qualitative features of the spectrum may be noted.

(a) At very high temperature (kT_4 in Figure 10.1) there is perfect gas behavior. Therefore the uppermost part of the spectrum is well approximated by states of an ideal gas.

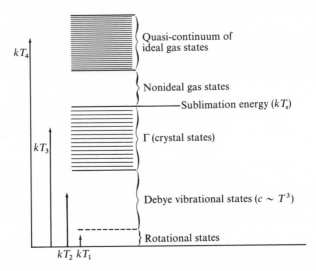

FIGURE 10.1
Spectrum of system that may exist in either gas or solid phase.

(b) Near absolute zero the hydrogen is frozen into a solid. It follows that the lowest part of the spectrum (kT_1 in Figure 10.1) contains rotational states with a level spacing determined by $\hbar^2 J(J + 1)/2I$ where I is the moment of inertia of the macroscopic crystal. The universal Debye spectrum, independent of crystal structure and leading to the T^3 law, becomes dominant at slightly higher but still very low temperatures (kT_2).

(c) At higher temperatures (kT_3) but still below the sublimation temperature, one observes the detailed structure of the crystal. In this intermediate energy range one finds states (Γ) which have the crystallographic symmetry of solid hydrogen.

(d) The Hamiltonian has complete rotational and translational symmetry. Consequently the set of all eigenstates of the Hamiltonian is a sum of irreducible representations of the rotation-translation group. However, a particular set of states, like Γ, may have less symmetry, and in particular those states that describe the crystal will have the appropriate crystallographic symmetry [1].

(e) The dimensionality of the representation (the degeneracy) is very important. In general the degeneracy decreases as the energy is decreased; the lowest level is nondegenerate, while levels of the gas are highly degenerate.

On the basis of these qualitative remarks we may now illustrate the sublimation of a solid by a very crude model (Figure 10.2). Idealize the spectrum by grouping all the gas states together (energy E_2, degeneracy g_2) and all the solid states together (energy E_1, degeneracy g_1) and assume, as already remarked, that

$$M = g_2/g_1 \gg 1. \tag{1.1}$$

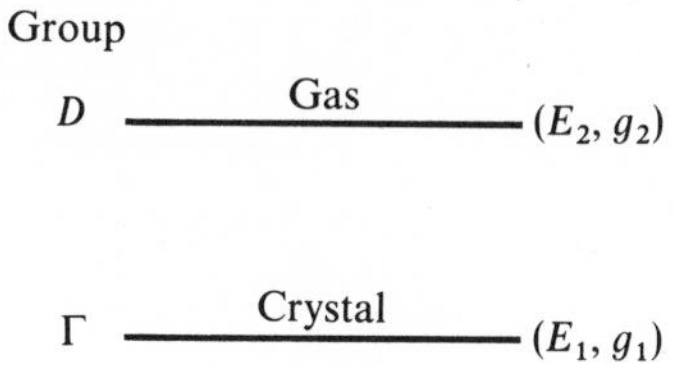

Group

D ——————— Gas ——————— (E_2, g_2)

FIGURE 10.2
Simple model of a two-phase system.

Γ ——————— Crystal ——————— (E_1, g_1)

In this model $E_2 - E_1 = l$ will be called the latent heat. Then

$$U(T) = \frac{g_1 E_1 e^{-E_1/kT} + g_2 E_2 e^{-E_2/kT}}{g_1 e^{-E_1/kT} + g_2 e^{-E_2/kT}} \tag{1.2}$$

$$= \frac{E_1 + M E_2 e^{-l/kT}}{1 + M e^{-l/kT}}, \tag{1.3}$$

and

$$U(0) = E_1 \tag{1.4a}$$

$$U(\infty) = \frac{E_1 + M E_2}{1 + M} \cong E_2. \tag{1.4b}$$

Define a transition temperature (T^*) by

$$U(T^*) = \frac{1}{2}(E_1 + E_2). \tag{1.5}$$

Then

$$M e^{-l/kT^*} = 1$$

and

$$kT^* = \frac{l}{\ln M}. \tag{1.6}$$

The entropy change at the transition is then

$$\frac{l}{T^*} = k \ln M = k \ln g_2 - k \ln g_1$$

$$= S_2 - S_1. \tag{1.7}$$

The width of the transition may be defined by

$$\delta T = T_g - T_s \tag{1.8}$$

where T_g and T_s are shown in Figure 10.3 and are necessarily somewhat arbitrary. For example, one may choose T_g and T_s to be determined as follows:

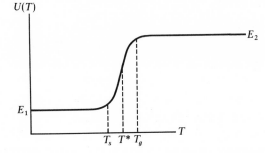

FIGURE 10.3
Energy as function of temperature. (Sublimation temperature
is T^*. Width of transition is
$T_g - T_s$.)

$$U(T_g) = \frac{E_1 + yE_2}{1 + y} \qquad (1.9a)$$

$$U(T_s) = \frac{E_1 + y^{-1}E_2}{1 + y^{-1}} \qquad (1.9b)$$

where the system is 90% solid at T_s and 90% gas at T_g if $y = 9$.
Then by (1.3) and (1.9)

$$\frac{\delta T}{T} \simeq \frac{\ln y^2}{\ln M}. \qquad (1.10)$$

This last relation shows that the transition is sharp if there is a large increase in the number of states between the one phase and the other. Since the degeneracy depends on the volume of the system, this relation implies a connection between the sharpness of the transition and the volume of the system. According to (1.7) and (1.10), $\delta T/T \sim (S_2 - S_1)^{-1}$ where $S_2 - S_1$ is proportional to the mass of the system. Therefore $\delta T/T \longrightarrow 0$ as the system becomes very large. It turns out that one may expect perfectly sharp transitions only in infinitely large systems.

Let us next proceed to a more detailed and realistic discussion.

10.2 The Many-body Problem and the Nonideal Gas

The Hamiltonian of the many-body problem is invariant under rotations and translations, and may be written as follows:

$$H_N = \sum_i p_i^2/2m_i + \sum_{i<j} V_{ij}. \qquad (2.1)$$

To discuss the condensation of a gas into a liquid, it is enough to work essentially at the classical level. Then the grand partition function is, according to equation (1.12) in Chapter 9,

$$Q = \sum_N Q_N \zeta^N \qquad (2.2)$$

where

$$Q_N = \frac{1}{h^{3N}} \frac{1}{N!} \int \cdots \int e^{-H_N/kT} \, d\mathbf{x}_1 \cdots d\mathbf{x}_N \, d\mathbf{p}_1 \cdots d\mathbf{p}_N. \qquad (2.3)$$

Spin and velocity dependent forces have been neglected in H_N. Then the potential V_{ij} between molecules i and j depends only on the separation $|\mathbf{x}_i - \mathbf{x}_j|$, and

$$Q_N = Q_{Np} \, Q_{Nx} \qquad (2.4)$$

where

$$Q_{Np} = \frac{1}{h^{3N}} \frac{1}{N!} \int \cdots \int e^{-\Sigma p_i^2/2mkT} \, d\mathbf{p}_1 \cdots d\mathbf{p}_N \qquad (2.4a)$$

$$Q_{Nx} = \int \cdots \int e^{-\Sigma V_{ij}/kT} \, d\mathbf{x}_1 \cdots d\mathbf{x}_N \qquad (2.4b)$$

since H_N is composed additively of parts that depend on the momenta only and the coordinates only. The momentum integral further simplifies

$$Q_{N_p} = \frac{Q_p^{\,N}}{N!}. \qquad (2.5)$$

Here

$$
\begin{aligned}
Q_p &= \frac{1}{h^3} \int e^{-p^2/2mkT} \, d\mathbf{p} \\
&= \frac{(2\pi mkT)^{3/2}}{h^3} = \frac{1}{\lambda^3}
\end{aligned}
\qquad (2.5a)
$$

where λ is called the thermal wavelength [see (2.24a), Chapter 8]. Hence

$$Q = \sum_N Q_{Nx} \frac{(Q_p \zeta)^N}{N!}. \qquad (2.6)$$

Let the expansions of the pressure and density deduced from (2.6) [according to equations (3.4) and (3.5) of Chapter 9] be

$$\frac{P}{kT} = \frac{1}{\lambda^3} \sum_l b_l \, \zeta^l \qquad (2.7)$$

$$\rho = \frac{1}{\lambda^3} \sum_l l \, b_l \, \zeta^l \qquad (2.8)$$

These two equations give the equation of state in parametric form; this is the cluster expansion and the b_l are called (reducible) cluster integrals [2], because they are associated with molecular clusters. The b_l depend in general on the volume (V) of the container, but it may be shown that in the vapor phase the preceding equations may be approximated by the following:

$$\frac{P}{kT} = \frac{1}{\lambda^3} \sum_l b_l \, \zeta^l \tag{2.7a}$$

$$\rho = \frac{1}{\lambda^3} \sum_l l \, b_l \, \zeta^l \tag{2.8a}$$

where

$$b_l(T) = \lim_{V \to \infty} b_l(T, V). \tag{2.9}$$

Under the same conditions one may also expand in $1/v$ and obtain the virial expansion [2]

$$\frac{Pv}{kT} = \sum_{l=1}^{\infty} a_l(T) \, (\lambda^3/v)^{l-1} \tag{2.10}$$

where v is the specific volume. The van der Waals equation, for example, may be regarded as an approximation to the complete virial expansion.

The a_l may be determined empirically, and the b_l may be expressed in terms of them [3]. Hence the b_l may also be regarded as measurable.

The difficulty with equations (2.7a) and (2.8a), however, is that these series cannot be used to describe the liquid phase since they do not converge outside of the gas phase. Nevertheless it will be seen in Section 10.7 how the cluster integrals may be related to the condensation. Before continuing the discussion we now turn to a simpler example.

10.3   Ferromagnetic Transitions and the Ising Model

A gaseous condensation is a first order phase change and relatively complex. Higher order transitions are smoother and perhaps mathematically more tractable. In particular a ferromagnetic transition is simpler than other phase changes insofar as we may regard the spin variable as discrete; but even here exact results have been obtained for only an idealized system, namely the Ising model. This model may be motivated as follows.

As a consequence of the exclusion principle, the fundamental Coulomb interaction between two electrons generates a spin dependent interaction of the form

$$J \, \sigma_1 \, \sigma_2 \tag{3.1}$$

where J is the exchange integral; J is of the order of magnitude of the Coulomb interaction, and leads to ferromagnetism when it is negative. The Ising model truncates this interaction by replacing (3.1) with

$$J \sigma_{1z} \sigma_{2z}. \tag{3.2}$$

Then the problem becomes classical, but even after this simplification the three dimensional case has still escaped an exact solution.

As the simplest interesting example of a solvable Ising magnet, consider a two dimensional square lattice of localized spins with the following properties. Each spin variable may take on only two values, say $+1$ and -1. There is an interaction only between nearest neighbors, and the interaction energy takes on only two values: $+J$ when the spins are antiparallel and $-J$ when they are parallel. Such a system is complicated enough to show a Curie point, but simple enough to be exactly solvable. (The one dimensional case is also solvable but there are not enough neighbors to produce a Curie transition.)

The following exact solution in the absence of a magnetic field was found by Onsager [4]:

$$\lim_{N \to \infty} \left(\frac{1}{N} \ln Q \right) = \ln \left(2 \cosh \frac{2J}{kT} \right) + \frac{1}{\pi} \int_0^{\pi/2} \ln \frac{1}{2} [1 + (1 - K^2 \sin^2 \omega)^{1/2}] \, d\omega \tag{3.3}$$

where

$$K = \frac{2 \sinh \dfrac{2J}{kT}}{\cosh^2 \dfrac{2J}{kT}}. \tag{3.3a}$$

The Curie point corresponds to

$$K = 1 \tag{3.4}$$

or

$$J/kT_c = 0.4407. \tag{3.4a}$$

At this point the specific heat becomes logarithmically divergent (Figure 10.4).

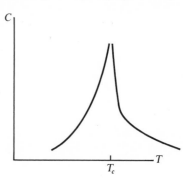

FIGURE 10.4
Curie transition for the Ising model (specific heat vs. temperature).

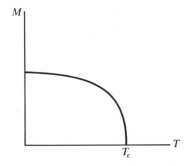

FIGURE 10.5
Spontaneous magnetization of the Ising magnet.

Although the exact behavior of the Ising model as a function of the magnetic field remains an unsolved problem, the spontaneous magnetization has been found by Yang [5]. It is

$$M = 0 \qquad\qquad T > T_c$$

$$M = N\mu \left[\frac{\cosh^2 2J/kT}{\sinh^4 2J/kT} (\sinh^2 2J/kT - 1) \right]^{1/8} \qquad T \leq T_c \qquad (3.5)$$

and is shown in Figure 10.5.

The Ising model remains the most important mathematical basis for our conjectures about the nature of classical phase transitions.

We now go on to a discussion of phase changes of the first kind which was suggested by the Ising problem. To do this it is first necessary to summarize some thermodynamic relations that hold for finite systems.

10.4 Equation of State of Finite Systems

We have seen how the equations of state may be formulated in the following way:

$$\frac{P}{kT} = \frac{1}{V} \ln Q \qquad (4.1)$$

$$\rho = \frac{1}{V} \frac{\partial \ln Q}{\partial \ln \zeta} \qquad (4.2)$$

$$u = \frac{1}{V} \frac{\partial \ln Q}{\partial \ln \chi} \qquad (4.3)$$

where ρ and u are the mass and energy densities respectively. See equations (3.4), (3.5), and (3.6) of Chapter 9.

By eliminating ζ between the first and second equations, one obtains the thermal equation of state; by eliminating ζ between the first and third one gets the caloric equation.

The corresponding formulas for fluctuations in these variables are given in Chapter 9, equation (4.3a). In particular we have found

$$\frac{\langle(\Delta N)^2\rangle}{N^2} = \frac{kT}{(\partial P/\partial\rho)_T}\frac{1}{N}. \tag{4.4}$$

The fractional fluctuation, which therefore varies as $N^{-1/2}$, is large for small systems and generally negligible for large systems. The limit in which N becomes very large will be referred to as the *thermodynamic limit*.

We shall discuss the equations of state in this limit, e.g.,

$$\frac{P}{kT} = \lim_{V\to\infty}\left(\frac{1}{V}\ln Q\right) \tag{4.5}$$

$$\rho = \lim_{V\to\infty}\left(\frac{1}{V}\frac{\partial\ln Q}{\partial\ln\zeta}\right) \tag{4.6}$$

for the thermal equation.

Although the fractional fluctuation in general vanishes for very large systems, it follows from (4.4) that if the isotherms become flat so that $(\partial P/\partial\rho)_T$ vanishes, then in these situations the fluctuations are still large. These fluctuations are naturally interpreted in terms of the clustering or formation of droplets which takes place when there is a change of phase.

Before going on to the discussion of the thermodynamic limit, we shall make some remarks on the formulas (4.1)–(4.3) which hold for finite V. We find

$$\rho = \frac{\partial}{\partial\ln\zeta}\left(\frac{P}{kT}\right) \tag{4.7}$$

$$u = \frac{\partial}{\partial\ln\chi}\left(\frac{P}{kT}\right). \tag{4.8}$$

It follows from these equations that P/kT is a monotonic increasing function of ζ and χ since ρ and u are positive.

We also find

$$\frac{\partial}{\partial\ln\zeta}\rho = \frac{\partial^2}{\partial(\ln\zeta)^2}\left(\frac{P}{kT}\right) = \frac{1}{V}\frac{\partial^2}{\partial(\ln\zeta)^2}\ln Q = \frac{\langle(\Delta N)^2\rangle}{V} > 0 \tag{4.9}$$

$$\frac{\partial}{\partial\ln\chi}u = \frac{\partial^2}{\partial(\ln\chi)^2}\left(\frac{P}{kT}\right) = \frac{1}{V}\frac{\partial^2}{\partial(\ln\chi)^2}\ln Q = \frac{\langle(\Delta E)^2\rangle}{V} > 0. \tag{4.10}$$

Therefore ρ and u are also monotonic increasing functions of ζ and χ respectively.

Let us now assume that the potential between the molecules is attractive except at distances less than a characteristic separation where it becomes infinitely repulsive. Since the molecules then behave at high densities like hard spheres, there is a maximum number (M) that may be packed into a box of volume V.

There is consequently an upper limit to ρ, namely, the density of closest packing, and therefore also, by equation (4.7), P must be a continuous function of ζ.

The grand partition function is now

$$Q = \sum_{1}^{M} Q_N \, \zeta^N \tag{4.11}$$

so that Q is a polynomial in ζ of degree M with positive coefficients Q_N. This polynomial has no roots on the positive ζ-axis and is an analytic function of ζ in the finite ζ-plane.

We may also write

$$Q(\beta, \alpha) = \iint e^{\beta E} \, e^{\alpha N} \, \rho(E, N) \, dE \, dN$$

$$\beta = -\frac{1}{kT} \qquad \alpha = \frac{\mu}{kT} \tag{4.12}$$

where

$$\rho(E, N) = 0 \quad \text{if} \quad E < 0 \quad \text{or} \quad N < 0. \tag{4.12a}$$

$Q(\beta, \alpha)$ is therefore always an analytic function in the left-half (β, α)-plane. However, if

$$\rho(E, N) = 0 \quad \text{for} \quad N > M \tag{4.12b}$$

as in the present situation, then $Q(\beta, \alpha)$ is analytic in the right-half α-plane as well; just as an upper limit to the energy spectrum permits an extension of $(Q(\alpha, \beta))$ into the region of negative temperature [6].

10.5   The Thermodynamic Limit and Phase Transitions [7, 8, 9]

As the volume of the box is increased, the roots of Q move about in the complex ζ-plane and their number increases. Their distribution in the limit $V \longrightarrow \infty$ gives the complete analytic behavior of the thermodynamic functions in the ζ-plane.

The essential results are that the limits of P and ρ exist and are monotonic increasing functions of ζ. While P is always continuous, ρ and u have discontinuities which depend, according to (4.9) and (4.10), on fluctuations in the particle and energy density. Let (ζ_1, χ_1) and (ζ_2, χ_2) correspond to states of phase 1 and phase 2. Then by (4.9) and (4.10)

$$\rho(2) - \rho(1) = \int_1^2 \frac{\langle (\Delta N)^2 \rangle}{V} \, d(\ln \zeta) \tag{5.1}$$

$$u(2) - u(1) = \int_1^2 \frac{\langle (\Delta E)^2 \rangle}{V} \, d(\ln \chi). \tag{5.2}$$

The discontinuities at the phase transition, where $\mu_2 = \mu_1$, $T_2 = T_1$ and $P_2 = P_1$, are

$$\lim_{\mu_2 \to \mu_1} V \int_1^2 \langle (\Delta \rho)^2 \rangle \, d(\ln \zeta) \tag{5.1a}$$

and

$$\lim_{T_2 \to T_1} V \int_1^2 \langle (\Delta u)^2 \rangle \, d(\ln \chi). \tag{5.2a}$$

The behavior of P/kT and ρ in the thermodynamic limit are controlled by the following two theorems, proved by Yang and Lee [7].

Theorem 1

For all positive real values of ζ, $V^{-1} \ln Q_V$ approaches, as $V \longrightarrow \infty$, a limit that is independent of the shape of V. Furthermore, this limit is a continuous, monotonically increasing function of ζ.

Theorem 2

If, in the complex ζ-plane, a region R containing a segment of the positive real axis is always free of roots of the grand partition function, then in this region as $V \longrightarrow \infty$ all the quantities

$$\frac{1}{V} \ln Q_V, \; \frac{\partial}{\partial \ln \zeta} \left(\frac{1}{V} \ln Q_V \right), \; \frac{\partial^2}{\partial (\ln \zeta)^2} \left(\frac{1}{V} \ln Q_V \right) \cdots$$

approach limits that are analytic with respect to ζ. Furthermore in R,

$$\lim_{V \to \infty} \frac{\partial}{\partial \ln \zeta} \left(\frac{1}{V} \ln Q_V \right) = \frac{\partial}{\partial \ln \zeta} \lim_{V \to \infty} \left(\frac{1}{V} \ln Q_V \right)$$

so that by (4.5) and (4.6),

$$\rho = \frac{\partial}{\partial \ln \zeta} \left(\frac{P}{kT} \right). \tag{5.3}$$

According to the first theorem, the pressure is always a continuous function

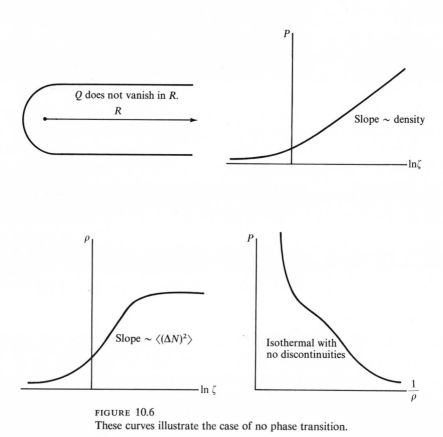

FIGURE 10.6
These curves illustrate the case of no phase transition.

of ln ζ, while according to the second theorem, the density acquires a discontinuity only if the limiting distribution of the roots cuts the positive real axis.

Examples

1. Suppose there exists a region R that contains the whole real axis and is free of roots. Figure 10.6 illustrates this case (which corresponds to the nonexistence of a phase transition).

2. Suppose, on the other hand, that in the limit $V \longrightarrow \infty$ the distribution of roots of Q_V cuts the positive real axis as shown in Figure 10.7. In this case there are two phase transitions. In general there is a phase transition whenever the distribution of roots cuts the positive real axis in the thermodynamic limit.

10.6 Lattice Gas and Phase Transitions

It is not possible to make a calculation leading from the exact Hamiltonian of a real physical system to the limiting distribution of the roots of the grand partition function. However, the theory can be carried through for certain model Hamiltonians which exhibit realistic qualitative features.

In particular we shall consider the lattice gas; in this model the molecules may be situated only at the discrete points of a lattice. The following theorem has been proved [9].

Theorem 3

If the interaction u between two gas molecules is such that $u = +\infty$ when two molecules occupy the same lattice site and $u \leq 0$ otherwise, then all the zeros of the grand partition function lie on a circle in the complex ζ-plane

In order to discuss the lattice gas we shall assume this theorem and write [10]

$$Q_V = \prod_{i=1}^{M} (\zeta - \zeta_i) \qquad (6.1)$$

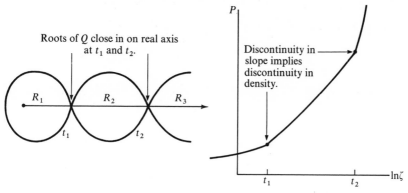

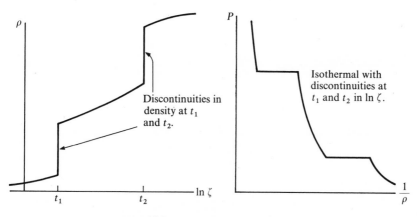

FIGURE 10.7
These curves illustrate two phase transitions.

where ζ_i are the M roots of Q_V and lie on a circle. Then

$$\frac{P}{kT} = \frac{1}{V} \ln Q_V \tag{6.2}$$

$$= \frac{1}{V} \sum_i \ln (\zeta - \zeta_i)$$

$$= \int_0^{2\pi} \ln (\zeta - e^{i\theta}) \left[\frac{g_V(\theta)}{V} \right] d\theta \tag{6.3}$$

where $g_V(\theta) \, d\theta$ is the number of roots in the interval $(\theta, \theta + d\theta)$ and for simplicity we have chosen the unit circle. The corresponding expression for ρ_V is

$$\rho_V = \frac{\partial}{\partial \ln \zeta} \left(\frac{P}{kT} \right). \tag{6.4}$$

We know that there are no real positive roots for any finite V and that therefore

$$g_V(0) = 0. \tag{6.5}$$

However, as V becomes infinite, the distribution of roots may approach the real axis so that the limit

$$\lim_{\theta \to 0} \lim_{V \to \infty} \left[\frac{g_V(\theta)}{V} \right] = g(0) \tag{6.6}$$

does not vanish. But if this does happen, then according to the general theory there should be a discontinuity in ρ at this point. That such a discontinuity does exist may be seen very simply in terms of an electrostatic analogue [9]; for the electrostatic potential (ϕ) due to an infinite cylinder, of unit radius, perpendicular to the ζ-plane and carrying a surface charge density, $[g_V(\theta)]/V$ is, according to (6.3),

$$\phi = -\frac{2}{V} \int_0^{\pi} \ln (\zeta^2 - 2\zeta \cos\theta + 1) \, g_V(\theta) \, d\theta$$

$$= -\frac{2P}{kT}. \tag{6.7}$$

The corresponding electric field on the real axis is, by (6.4),

$$E = \frac{2\rho_V}{\zeta}. \tag{6.8}$$

For finite V there is a gap in the charge distribution at $\theta = 0$. If the distribution of roots does not pinch the real axis in the limit $V = \infty$, then the electric field remains continuous as one passes through this gap from the

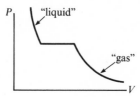

FIGURE 10.8
Isothermal for lattice gas deduced
from the Ising model.

interior to the exterior of the cylinder. On the other hand, if $g(0) \neq 0$, then the gap is filled with a single layer which gives rise to a discontinuity in the electric field of amount

$$\Delta E = 4\pi \, g(0), \tag{6.9}$$

and therefore the discontinuity in density, if there is a phase change, is

$$\Delta\rho = 2\pi \, g(0). \tag{6.10}$$

The preceding discussion gives a simple description of phase transitions provided that the distribution of roots does cut the real axis in the thermodynamic limit. Although this conjectured behavior of the limiting distribution has not been demonstrated in general, it has been established in an important special case, the two dimensional Ising model; in this case there is an exact proof that the roots do cut the real axis in the limit, although even in this case the complete $g(\theta)$ is not known [9]. The two dimensional lattice gas is mathematically equivalent to the Ising model and from the exact solutions of the latter, one may deduce for example the exact isotherms of the former [9]; below the critical point the flat portion (Figure 10.8) of the isotherm is rigorously predicted.

10.7 Connection with the Theory of the Nonideal Gas

The description of a phase transition that emerges from the Yang-Lee theory is very attractive in its essentials, but in principle at least one would like to deduce the distribution of the zeros of the grand partition function from the atomic Hamiltonian or the forces between the molecules. Although it has not yet been possible to proceed very far in this direction, there are some quite simple results relating $g(\theta)$ to the cluster integrals and virial expansion.

In (2.7a) there is the following expansion of the pressure

$$\frac{P}{kT} = \frac{1}{\lambda^3} \sum_l b_l \, \zeta^l \tag{7.1}$$

where the b_l are the reducible cluster integrals. One may now compare this with the representation of P/kT in terms of $g(\theta)$:

$$\frac{P}{kT} = \int_0^{2\pi} \ln{(\zeta - e^{i\theta})} \, g(\theta) \, d\theta \tag{7.2}$$

according to (6.3), where $g(\theta) = \lim_{V \to \infty} (g_V(\theta)/V)$. By differentiation we find

$$\frac{d^n}{d\zeta^n} \left(\frac{P}{kT}\right) = -\int_0^{2\pi} \frac{(n-1)!}{(e^{i\theta} - \zeta)^n} \, g(\theta) \, d\theta$$

and

$$\left[\frac{1}{n!} \frac{d^n}{d\zeta^n} \left(\frac{P}{kT}\right)\right]_{\zeta=0} = -\frac{1}{n} \int_0^{2\pi} e^{-in\theta} \, g(\theta) \, d\theta. \tag{7.3}$$

Comparing (7.3) with the cluster expansion (7.1), we find

$$\lambda^{-3} \beta_l = -\frac{1}{l} \int_0^{2\pi} e^{-il\theta} \, g(\theta) \, d\theta. \tag{7.4}$$

Therefore the reducible cluster integrals β_l are the Fourier coefficients of $g(\theta)$.

Since the virial coefficients a_l, and hence the β_l, may be determined empirically, it is then in principle possible to determine the distribution function $g(\theta)$.

10.8   Phase Transitions of a Quantum Fluid

The essential feature of a quantal phase transition appears to be the macroscopic occupation of a single quantum mechanical state. If this occupation were complete, the corresponding thermodynamic state would be pure and one would have $\rho^2 = \rho$; but this condition holds only at $T = 0$. Since a quantum phase change may occur at fairly high temperatures, the desired criterion for the existence of such a transition must be much less demanding than the requirement that $\rho^2 = \rho$. Since it is not clear how to approach this question in a general way, let us mention some of the main points about the known transitions.

The onset of superconductivity is associated with the appearance of an effective energy gap, and similarly the existence of superfluidity is associated with a paucity of low lying excitations in helium II.

On the other hand, the two cases appear to be unlike in that the Einstein-Bose condensation occurs even if there is no interaction between the particles.

In the two cases the mathematical models are similar, namely, an ideal gas with a weak interaction, and precisely the Hamiltonian already given in Chapter 9, equation (6.30):

$$H = \sum_k \bar{a}(k) \, a(k) \, \epsilon(k) + \frac{1}{2V} \sum_{k,l,n} \phi(n) \, \bar{a}(k) \, \bar{a}(l) \, a(l-n) \, a(k+n) \tag{8.1}$$

where

$$\phi(\mathbf{n}) = \lambda \int u(\mathbf{r}) \, e^{i\mathbf{n}\mathbf{r}} \, d\mathbf{r}. \tag{8.1a}$$

This kind of Hamiltonian is used not only to describe the superfluid [11] but also to describe the superconductor [12]. Of course, in the Fermi-Dirac case, spin variables must be introduced, but they do not change the formal appearance of the interaction.

The same Hamiltonian may be expressed in configuration space with the aid of field operators [Chapter 9, equation (6.27b)]:

$$H = \int \bar{\psi}(\mathbf{x}) \left(-\frac{\hbar^2 \nabla^2}{2m} \right) \psi(\mathbf{x}) \, d\mathbf{x} + \frac{\lambda}{2} \iint d\mathbf{x} \, d\mathbf{y} \, u(\mathbf{x} - \mathbf{y}) \, \bar{\psi}(\mathbf{x}) \, \bar{\psi}(\mathbf{y}) \, \psi(\mathbf{x}) \, \psi(\mathbf{y}).$$
$$\tag{8.2}$$

The equation of motion which follows from (8.2) is both nonlocal and nonlinear. However, if we take $u(\mathbf{x})$ to be short range and put

$$g = \lambda \int u(\mathbf{x}) \, d\mathbf{x},$$

then the equation of motion of $\psi(\mathbf{x})$ becomes [13]

$$i\hbar \frac{\partial \psi}{\partial t} = -\frac{\hbar^2}{2m} \nabla^2 \psi + g(\bar{\psi}\psi)\psi. \tag{8.3}$$

If one now writes

$$\psi = a_0 + \theta \tag{8.4}$$

and regards a_0 as macroscopic and classical, one obtains the classical equation

$$i\hbar \frac{\partial a_0}{\partial t} = -\frac{\hbar^2 \nabla^2}{2m} a_0 + g|a_0|^2 \, a_0. \tag{8.5}$$

This is the Pitaevskii equation [13]. A similar kind of equation (Ginsburg-Landau) holds in the superconductive case [14]. These equations have been used successfully in phenomenological theories [15].

We now turn to a quantum mechanical treatment of (8.1).

10.9   The B.C.S. Solution

The B.C.S. theory assumes that the superconductive electronic system behaves like an ideal, completely degenerate Fermi gas, except for a very weak attraction between pairs of electrons that have opposite spins and momenta, and are also very near the top of the Fermi sea. The effective or reduced

Hamiltonian is thus of type (8.1) but will be written in a slightly more convenient notation [12]:

$$H_{reduced} = \sum_{k,s} \epsilon_k\, n_{ks} + \sum_{lk} V_{lk}\, \bar{b}_l\, b_k \tag{9.1}$$

where ϵ_k is the Fermi energy corresponding to the momentum $\mathbf{k}$, and n_{ks} is the number of electrons with momentum $\mathbf{k}$, and spin s. The first term then represents the kinetic energy and the second term represents the pairing interaction. Here

$$V_{lk} = (l - l|V|k - k) \tag{9.2}$$

$$b_k = a_{k\uparrow}\, a_{-k\downarrow}. \tag{9.3}$$

The $(a_{k\uparrow}, \bar{a}_{k\uparrow})$ are absorption and emission operators for the single electrons $(\mathbf{k}\uparrow)$, and therefore $\bar{b}_k$ creates a pair of electrons in the single particle states $\mathbf{k}\uparrow$ and $-\mathbf{k}\downarrow$. Thus both spins and currents are paired to cancel. The matrix element for scattering vanishes unless $\mathbf{k} + \mathbf{l} = \mathbf{p} + \mathbf{q} = 0$ and is determined by the function V_{kl}. According to the simple B.C.S. model, V_{kl} also vanishes unless $\mathbf{k}$ and $\mathbf{l}$ are both very close to the Fermi surface.

If this Hamiltonian does indeed correspond to the actual physical situation, then the state function $|S\rangle$ which minimizes $\langle S|H|S\rangle$ for a given number of particles is by definition the superconductive ground state. Then

$$\delta \langle S|H_{reduced} - \mu N|S\rangle = 0 \tag{9.4}$$

where μ is a Lagrangian multiplier arising from the constraint on the total number of particles, and having the interpretation of chemical potential. To obtain the B.C.S. solution one puts $\mu = 0$; this assumption is of course equivalent to dropping the constraint on the number of particles [12].

The usual variational procedure is now to assume that $|S\rangle$ lies in a particular function space and depends on certain parameters (ϕ_k). When these parameters are varied to satisfy (9.4), one obtains a state $|S\rangle$ which, according to the Ritz variational principle, is the best choice for the ground state lying in the original function space and determined by $H_{reduced}$ (subject to the given constraint on N).

Let $|F\rangle$ be the ground state of a Fermi gas. Then the B.C.S. ansatz for $|S\rangle$ is

$$|S\rangle = \prod_{k}^{|\epsilon_k| < \omega} (\cos \phi_k + \sin \phi_k\, \bar{b}_k)|F\rangle. \tag{9.5}$$

Here ϵ_k is the energy measured from the top of the Fermi sea; therefore the product runs over only those momentum states in a narrow band which is centered at the top of the Fermi sea and has width 2ω—this is the range in which V_{kl} does not vanish.

If one now puts $|S\rangle$ in (9.4) and varies with respect to the ϕ_k, one obtains an integral equation for ϕ_k. To solve this equation in an elementary way, put

$$V_{kl} = -J \ (= \text{constant}) \tag{9.6}$$

where V_{kl} does not vanish. Then one obtains [12]

$$\tan 2\phi_k = \frac{\Delta}{\epsilon_k} \tag{9.7}$$

where Δ is the energy gap and is

$$\Delta \cong 2\omega \, e^{-1/[N(0)J]} \tag{9.8}$$

where $N(0)$ is the density of states at the surface of the Fermi sea. Equations (9.7) and (9.8) determine ϕ_k and therefore the ground state in terms of known quantities [16].

Although the above solution has been described in terms of the variational method by which it was first obtained, it was later shown by Bogoliubov that the B.C.S. solution is in fact an exact solution for the given Hamiltonian in the limit of infinite volume [17].

One notes particularly the nonperturbative nature of the solution: Δ is not an analytic function of J near $J = 0$.

The B.C.S. theory is based on a truncated Hamiltonian which has not been obtained from first principles, but the theory has now been worked out in much detail and is in convincing agreement with experiment [18].

Although richer in physical content than the Ising model, the B.C.S. model appears to play a somewhat similar role, in that it arises as a soluble truncation of a much more complicated physical problem. The Ising and B.C.S. solutions provide the main source of mathematically exact information about classical and quantal phase transitions.

In both cases there is a fundamental spin correlation, which in the B.C.S. case is antimagnetic in the currents as well as the spins and leads to a diamagnetic instead of a ferromagnetic ground state.

10.10   Coherence in the Condensed State

Classical phase transitions may be described in terms of classical order parameters which measure the reorganization of the physical system at the transition. In the quantum case these order parameters should measure true quantum mechanical coherence. A method for describing quantal order will now be presented.

We have discussed the Fermi-Dirac condensation in terms of the B.C.S. theory and the Einstein-Bose condensation in terms of the free particle theory. There is also a detailed theory of the nonideal boson liquid, developed by Landau, Feynman, Lee and Yang, Bogoliubov, and others. As Penrose and Onsager [19] noted, however, the theories of superfluid systems in terms of nonideal E.B. or F.D. fluids do not appear to be very directly related to the simple E.B. condensation of an ideal gas. They therefore proposed a simple criterion for an E.B. condensation which is applicable to nonideal as well as to ideal systems, and provides a simpler and more unified view of a quantal condensation. This criterion is formulated in terms of reduced density matrices (Section 9.6) and has been generalized by Yang [20] to F.D. systems.

CONDENSATION OF IDEAL EINSTEIN-BOSE SYSTEM

The condition for the condensation of N noninteracting bosons may be formulated in terms of a reduced density matrix ρ_1.

Let the statistical matrix for the ensemble associated with an ideal gas be

$$\rho = \sum_{n_1, \ldots} |n_1, \ldots\rangle \, W(n_1, \ldots) \, \langle n_1, \ldots|. \tag{10.1}$$

Then by equation (6.18) of Chapter 9,

$$\langle k|\rho_1|l\rangle = \operatorname{Tr} \rho \, \bar{a}_l \, a_k \tag{10.2}$$

where l and k are two momentum states, and by (10.1)

$$\begin{aligned}
\langle k|\rho_1|l\rangle &= \sum_{\substack{n_1, \ldots \\ n_1', \ldots}} \langle n_1', \ldots|n_1, \ldots\rangle \, W(n_1, \ldots) \, \langle n_1, \ldots|\bar{a}_l \, a_k|n_1', \ldots\rangle \\
&= \delta_{lk} \, \langle k|\rho_1|k\rangle \\
&= \langle N_k\rangle \, \delta_{lk}
\end{aligned} \tag{10.3}$$

according to (6.17) of Chapter 9. Although ρ_1 is diagonal in the momentum representation, it is not diagonal in the coordinate representation:

$$\begin{aligned}
\langle \mathbf{x}|\rho_1|\mathbf{y}\rangle &= \sum_{k,l} \langle \mathbf{x}|\mathbf{k}\rangle \, \langle k|\rho_1|l\rangle \, \langle \mathbf{l}|\mathbf{y}\rangle \\
&= \frac{1}{V} \sum_k \langle N_k\rangle \, e^{i\mathbf{k}(\mathbf{x}-\mathbf{y})}.
\end{aligned} \tag{10.4}$$

Above the E.B. condensation temperature,

$$\lim_{|x-y|\to\infty} \langle \mathbf{x}|\rho_1|\mathbf{y}\rangle = 0. \tag{10.5}$$

Below the E.B. transition, the sum over $\mathbf{k}$ approaches the contribution from the ground state ($\mathbf{k}_0 = 0$) alone. Then if we put $\mathbf{k}_0 = 0$,

$$\lim_{|\mathbf{x}-\mathbf{y}|\to\infty} \langle \mathbf{x}|\rho_1|\mathbf{y}\rangle = \frac{n_0}{V} \tag{10.6}$$

where n_0 is the macroscopic population of the ground state and is proportional to V. Therefore

$$\lim_{|\mathbf{x}-\mathbf{y}|\to\infty} \langle \mathbf{x}|\rho_1|\mathbf{y}\rangle \neq 0 \tag{10.7}$$

below the condensation temperature. This condition describes the so-called off-diagonal long range order (ODLRO).

REFORMULATION OF CRITERION

Next consider a nonideal system. Let the eigenvectors of the matrix $\langle y|\rho_1|x\rangle$ be $\phi_n(x)$. Then

$$\int \langle y|\rho_1|x\rangle \, dx \, \phi_n(x) = \lambda_n \, \phi_n(y)$$

where x is an abbreviation for $(\mathbf{x}_1, \ldots, \mathbf{x}_N)$.

Assume that the $\phi_n(x)$ are complete. Then

$$\sum_n \phi_n(x) \, \bar{\phi}_n(y) = \delta(x - y)$$

and therefore

$$\langle x|\rho_1|y\rangle = \sum_n \lambda_n \, \phi_n(x) \, \bar{\phi}_n(y). \tag{10.8}$$

Now if there is one very large eigenvalue (λ_0), then

$$\langle x|\rho_1|y\rangle = \lambda_0 \, \phi_0(x) \, \bar{\phi}_0(y) + \delta_1 (x, y) \tag{10.9}$$

where $\delta_1(x, y)$ is a remainder which becomes small in the limit $x - y \longrightarrow \infty$. The step from (10.8) to the limit of (10.9) is just the same as from (10.4) to (10.6). If we normalize so that $\phi_0(x) \sim 1/\sqrt{V}$, then if $\lambda_0 \sim N$, it follows that

$$\lim_{\mathbf{x}-\mathbf{y}\to\infty} \langle x|\rho_1|y\rangle \neq 0. \tag{10.10}$$

The criterion for (10.10) is that there exist an eigenvalue of ρ_1 of the order of N.

The corresponding statement for an F.D. system is that the second reduced matrix have a macroscopic eigenvalue.

Yang now proposes that the criterion for an E.B. (F.D.) condensation is that $\rho_1(\rho_2)$ acquire an eigenvalue of the order of N [20].

SUPERCONDUCTIVE CONDENSATION OF B.C.S. SYSTEM

One may test this criterion on the B.C.S. model. In this case it is necessary to consider ρ_2 where

$$\langle kl|\rho_2|ij \rangle = \mathrm{Tr}\; a_k\, a_l\, \rho\, \bar{a}_j\, \bar{a}_i.$$

At $T = 0$, there is a pure state $|S\rangle$ and by equation (6.21), Chapter 9,

$$\langle kl|\rho_2|ij \rangle = \langle S|\bar{a}_j \bar{a}_i a_k a_l|S\rangle.$$

Since $|S\rangle$ is known from (9.7) and (9.8), it is a straightforward calculation to determine the matrix elements and the eigenvalues of ρ_2. One finds that the largest eigenvalue of ρ_2 is of the order of

$$N(0)\,\Delta$$

which is of the order of N as required [21]. (Here Δ is the energy gap and $N(0)$ is the density of states at the surface of the Fermi sea; see Section 10.9.)

Notes and References

1. As an illustration of this situation consider an irreducible representation of the rotation group, say $D(2)$; the basis functions are then all the five states of angular momentum $(J = 2)$. These may be decomposed as follows:

$$D(2) = \Gamma_5 + \Gamma_3$$
$$(5 = 3 + 2)$$

   where Γ_5 and Γ_3 are two different irreducible representations of the cubic group (of dimensions 3 and 2).
2. For the original literature on the cluster expansion of Ursell and Mayer, see J. E. Mayer and M. G. Mayer, *Statistical Mechanics*, Chapter 13 (Wiley, New York, 1940).
   See also, K. Huang, *Statistical Mechanics* (Wiley, New York, 1963).
3. K. Huang, reference [2], p. 303.
4. L. Onsager, *Phys. Rev.* 65: 117 (1944). See also, for example, K. Huang, reference [2], p. 370, and G. Wannier, *Statistical Physics*, p. 338 (Wiley, New York, 1966).
5. C. N. Yang, *Phys. Rev.* 85: 809 (1952).
6. See note [4] of Chapter 7.
7. The first important investigation of this limit was made by L. van Hove, *Physica* 15: 951 (1949). He showed

$$\frac{\partial}{\partial v}\lim_{N\to\infty}\left[\frac{1}{N}\ln Q_N(vN)\right] = \lim_{V\to\infty}\left[\frac{1}{V}\ln \sum_0^\infty \zeta^n Q_n(V)\right].$$

   The left side of this equation is the pressure computed from the canonical ensemble, while the right side is the pressure derived from the grand canonical ensemble. One way of stating van Hove's theorem (Huang) is the following: the equation of state derived from the canonical ensemble agrees with that derived from the grand canonical ensemble in the thermodynamic limit. The discussion reproduced here follows.

8. C. N. Yang and T. D. Lee, *Phys. Rev.* 87: 404 (1952).
9. C. N. Yang and T. D. Lee, *Phys. Rev.* 87: 410 (1952).
10. The polynomial representing Q_v differs from (6.1) by a constant which is not important in the present context.
11. L. D. Landau and E. M. Lifshitz, *Statistical Physics*, p. 241 (Pergamon, London, 1958). I. M. Khalatnikov, *Introduction to the Theory of Superfluidity*, Chapter 5 (Benjamin, New York, 1965).
12. J. R. Schrieffer, *Theory of Superconductivity*, p. 36 (Benjamin, New York, 1964).

    The quantum theory of superconductivity was long an outstanding problem in theoretical physics. For the discovery of the effective Hamiltonian and the physical approximations needed to discuss it see N. N. Bogoliubov, *The Theory of Superconductivity* (Gordon and Breach, New York, 1962).

    The sketch given in section (10.9) is algebraically simple but implies certain physical approximations which are not discussed here. For some discussion of these approximations see the above books of Bogoliubov and Schrieffer.

13. L. P. Pitaevskii, *Soviet Physics—J.E.T.P.* (U.S.S.R.) 13: 451 (1961). The equation of motion following directly from (8.2) is

$$i\hbar \frac{\partial \psi}{\partial t} = [\psi, H]$$

$$= - \frac{\hbar^2}{2m} \nabla^2 \psi + \lambda \int dy \, \bar\psi(y) \, \psi(y) \, u(x - y) \, \psi(x).$$

    If $u(x)$ is now taken to be a δ-function or of short range, one gets (8.3).

14. Schrieffer, reference [12], p. 253. The Ginsburg-Landau-Gorkov nonlinear wave equation is

$$\left\{ \frac{1}{2m} \left[\nabla - i \frac{e^*}{c} A \right]^2 + \frac{1}{\lambda_G} \left[\left(1 - \frac{T}{T_c} \right) - \frac{2}{N} |\psi(r)|^2 \right] \right\} \psi(r) = 0$$

    where $e^* = 2e$. λ_G is the Gorkov parameter and the "wave function" $\psi(r) \sim \Delta(r)$, the energy gap.

15. P. G. de Gennes, *Superconductivity of Metals and Alloys* (Benjamin, New York, 1966).
16. The term "energy gap" has not been justified by the summary given here. See Schrieffer, reference [12].
17. N. N. Bogoliubov, *Physica* 26: 1 (1960).
    See also R. Haag, *Nuovo Cimento* 25: 287 (1962).
    N. M. Hugenholtz, "Reports on Progress" in *Physics* 28: 201 (1965).

    The essential point is that the equations of motion which follow from (8.2) are only apparently nonlinear in the field operators in the limit of infinite volume. That is because the nonlinear term contains the operator

$$\Delta(x) = \frac{1}{2V} \int dy \, dz \, u(x, y - z) \, \psi(z) \, \psi(y)$$

and this operator is in fact a *c*-number in the limit $V \longrightarrow \infty$, because of the finite range of $u(x)$. Since $\Delta(x)$ is a *c*-number, the equations of motion are linear in the annihilation and creation operators and may be solved exactly—most conveniently by the so-called Bogoliubov transformation:

$$b_k = u_k a_k + v_k \bar{a}_{-k}$$

$$u_k^2 + v_k^2 = 1.$$

18. See Schrieffer, reference [12], and de Gennes, reference [15], for example.
19. O. Penrose and L. Onsager, *Phys. Rev.* 104: 576 (1956).
20. C. N. Yang, *Rev. Mod. Phys.* 34: 694 (1962).
21. M. Rensink, *Ann. Phys.* 44: 105 (1967).

11

The Approach to
Thermodynamic Equilibrium

11.1 Introduction

We have discussed equilibrium statistical mechanics as the basis for thermodynamics, but have not yet described the detailed mechanisms for establishing statistical equilibrium. In particular it has not been shown how a system that is prepared in a nearby state will most probably move toward equilibrium by the operation of the so-called transport processes. In order to discuss these mechanisms we shall first go over from the Gibbs picture (Γ-space), in which the representative point corresponds to the complete gas, to the Boltzmann picture (μ-space), in which there is a point in phase space for each molecule. The effects of collisions on the distribution of the molecules over phase space may then be examined.

In order to show the relation between the Gibbs and Boltzmann pictures, one may rewrite the formulas of Chapter 7, which were obtained by the Gibbs method in the sense that they are summed over states of the entire gas. It will then be shown how these formulas, after they are rewritten, may be interpreted in the Boltzmann language.

11.2 Population Numbers as Variables

A quantum mechanical state may be described by the population numbers $(n_1, n_2, \ldots)$, which are eigenvalues of the number operators. A thermodynamic state on the other hand is specified by the corresponding expectation values $(\bar{n}_1, \bar{n}_2, \ldots)$. It is now interesting to express the entropy in terms of these expectation values.

The Helmholtz free energy may be expressed in terms of ζ and the Boltzmann factors x_i by (2.18) of Chapter 8:

$$A = kT \left[n \ln \zeta \mp \sum_i g_i \ln (1 \pm \zeta x_i) \right]. \tag{2.1}$$

The x_i also determine the expectation values $\langle n_i \rangle$:

$$\langle n_i \rangle = \frac{g_i}{\zeta^{-1} x_i^{-1} \pm 1}.$$

Define

$$\nu_i = \langle n_i \rangle / g_i = \frac{1}{\zeta^{-1} x_i^{-1} \pm 1}$$

then

$$x_i = \frac{\zeta^{-1} \nu_i}{1 \mp \nu_i}. \tag{2.2}$$

Let us now write A in terms of the ν_i, the population variables. We have by (2.2)

$$n \ln \zeta + \sum_i \langle n_i \rangle \ln x_i = \sum_i \langle n_i \rangle \ln \frac{\nu_i}{1 \mp \nu_i}$$

or

$$n \ln \zeta = \frac{U}{kT} + \sum_i \langle n_i \rangle \ln \frac{\nu_i}{1 \mp \nu_i}.$$

By (2.1), (2.2), and the preceding equation,

$$\frac{A}{kT} = \frac{U}{kT} + \sum_i \langle n_i \rangle \ln \frac{\nu_i}{1 \mp \nu_i} \pm \sum_i g_i \ln (1 \mp \nu_i)$$

$$= \frac{U}{kT} + \sum_i g_i \left[\nu_i \ln \frac{\nu_i}{1 \mp \nu_i} \pm \ln (1 \mp \nu_i) \right]$$

$$= \frac{U}{kT} + \sum_i g_i [\nu_i \ln \nu_i \pm (1 \mp \nu_i) \ln (1 \mp \nu_i)]. \tag{2.3}$$

For the entropy, one then finds

$$S = -k \sum_i g_i [\nu_i \ln \nu_i \pm (1 \mp \nu_i) \ln (1 \mp \nu_i)], \tag{2.4}$$

and therefore

$$H = \sum_i g_i[\nu_i \ln \nu_i \pm (1 \mp \nu_i) \ln (1 \mp \nu_i)]. \tag{2.5}$$

In the Fermi-Dirac case the term $\nu_i \ln \nu_i$ can be attributed to the particles and the $(1 - \nu_i) \ln (1 - \nu_i)$ to the holes. In both the Fermi-Dirac and Einstein-Bose cases, if $\nu_i \ll 1$,

$$\bar{H} = \sum_i g_i[\nu_i \ln \nu_i - \nu_i]. \tag{2.6}$$

This is the classical limit where the density is so low that the average population of a cell is very small.

Instead of (2.5) one may also write

$$\bar{\bar{H}} = \sum_{\text{states}} [\nu_i \ln \nu_i \pm (1 \mp \nu_i) \ln (1 \mp \nu_i)] \tag{2.7}$$

where the sum is carried out over states instead of over energy levels.

The expression (2.7) for $\bar{\bar{H}}$ has just been obtained by rewriting the equilibrium formula (2.1). It will now be regarded as a definition of $\bar{\bar{H}}$ which is valid for nonequilibrium situations as well.

11.3   Alternative Derivation of the Entropy

The result (2.4) may also be obtained by counting the number (W) of microscopic states that correspond to a given macroscopic state. Then according to Boltzmann, the entropy is

$$S = k \ln W. \tag{3.1}$$

Consider an ideal gas and the distribution of its molecules in momentum space. Let the number of molecules having a given energy (ϵ) be n_ϵ and let the degeneracy of that energy level be g_ϵ (where the g_ϵ states differ in the direction of motion and in the spin). Let $W(n, g)$ be the total number of ways of distributing n molecules over the g cells having the same energy. Then the required number of microscopic states is

$$W = \prod_\epsilon W(n_\epsilon, g_\epsilon), \tag{3.2}$$

and therefore

$$S = k \sum_\epsilon \ln W(n_\epsilon, g_\epsilon). \tag{3.3}$$

The function $W(n, g)$ may be calculated by noting that it is the number of solutions of the equation

$$\sum_{i=1}^{g} n_i = n. \tag{3.4}$$

In the F.D. case, $n_i = 0, 1$, and in the E.B. case, n_i may be any integer. In the F.D. case, one may write

$$\underbrace{(1 + x)(1 + x) \cdots (1 + x)}_{g \text{ factors}} = (1 + x)^g \tag{3.5}$$

or

$$\sum_{n} W(n, g)x^n = (1 + x)^g.$$

Therefore

$$W(n, g) = \binom{g}{n}. \tag{3.6}$$

In the E.B. case, one writes

$$(1 + x + x^2 + \cdots)(1 + x + x^2 + \cdots) \cdots (1 + x + x^2 + \cdots)$$
$$= \left(\frac{1}{1 - x}\right)^g \tag{3.7}$$

or

$$\sum_{n} W(n, g)x^n = (1 - x)^{-g}.$$

Therefore

$$W(n, g) = \binom{g + n - 1}{n}. \tag{3.8}$$

By using these expressions for $W(n, g)$ and Sterling's formula, one finds (2.4).

11.4   Kinetic Description

The connection between statistical theory and thermodynamics has been based on the Gibbs method and on the Γ-space in which the total system is regarded as an irreducible unit. Following Pauli [1], we now consider the fundamental problem from the Boltzmann point of view in μ-space.

Consider a collision taking two particles from initial state (p_i, p_j) to final state (p_k, p_l). According to classical ideas the rate of such collisions is proportional to

$$(kl|A|ij)\nu_i\nu_j, \tag{4.1}$$

where the $(kl|A|ij)$ are certain numbers depending on the types of molecules and specifying the "collision cross-section." According to quantum mechanics we shall assume that the rate depends on the population of the final as well as the initial states in the following way:

$$\text{F.D.:} \qquad (1 - \nu_k)(1 - \nu_l)(kl|A|ij)\nu_i\nu_j \qquad (4.2)$$

$$\text{E.B.:} \qquad (1 + \nu_k)(1 + \nu_l)(kl|A|ij)\nu_i\nu_j. \qquad (4.3)$$

In the F.D. case the classical rate is decreased by the particles that already occupy the final state. In the E.B. case there is the opposite effect.

The rate of change of the population of one cell is then

$$\frac{d\nu_i}{dt} = \sum_{j,k,l} [(1 \mp \nu_i)(1 \mp \nu_j)(ij|A|kl)\nu_k\nu_l - (1 \mp \nu_k)(1 \mp \nu_l)(kl|A|ij)\nu_i\nu_j]. \qquad (4.4)$$

The coefficients $(ij|A|kl)$ have certain symmetries. They are symmetric in the first two and also in the last two indices:

$$(ij|A|kl) = (ji|A|kl)$$

$$(ij|A|kl) = (ij|A|lk). \qquad (4.5)$$

In addition it may be shown that the forward and backward rates are equal:

$$(ij|A|kl) = (kl|A|ij). \qquad (4.6)$$

This symmetry is called microscopic reversibility and may be demonstrated in quantum as well as classical theory provided that the dynamical equations are invariant under time reversal and the rotation-reflection group.

Combining these various symmetries we have

$$\frac{d\nu_i}{dt} = \sum_{j,k,l} (ij|A|kl)(ij|B|kl) \qquad (4.7)$$

where

$$(ij|B|kl) = (1 \mp \nu_i)(1 \mp \nu_j)\nu_k\nu_l - (1 \mp \nu_k)(1 \mp \nu_l)\nu_i\nu_j$$
$$= (ij|b|kl) - (kl|b|ij) \qquad (4.8)$$

where

$$(ij|b|kl) = (1 \mp \nu_i)(1 \mp \nu_j)(\nu_k\nu_l). \qquad (4.9)$$

Clearly,

$$(ij|B|kl) = -(kl|B|ij). \qquad (4.10)$$

11.5 Change of $\bar{\bar{H}}$ in Time

We may now calculate the rate of change of $\bar{\bar{H}}$ from (2.7) and (4.7):

$$\frac{d\bar{\bar{H}}}{dt} = \sum_i \frac{d\nu_i}{dt} \ln \frac{\nu_i}{1 \mp \nu_i}$$

$$= \sum_{i,j,k,l} (ij|A|kl)(ij|B|kl) f_i \qquad (5.1)$$

where

$$f_i = \ln \frac{\nu_i}{1 \mp \nu_i}. \qquad (5.1a)$$

By permuting the summed indices this expression may be written more symmetrically:

$$\frac{d\bar{\bar{H}}}{dt} = \frac{1}{4} \sum_{i,j,k,l} (ij|A|kl)(ij|B|kl)[f_i + f_j - f_k - f_l]. \qquad (5.2)$$

But

$$f_i + f_j - f_k - f_l = \ln \frac{\nu_i \nu_j (1 \mp \nu_k)(1 \mp \nu_l)}{\nu_k \nu_l (1 \mp \nu_i)(1 \mp \nu_j)}; \qquad (5.3)$$

therefore

$$\frac{dH}{dt} = -\sum_{i,j,k,l} (ij|A|kl)[(ij|b|kl) - (kl|b|ij)] \ln \frac{(ij|b|kl)}{(kl|b|ij)}. \qquad (5.4)$$

Putting

$$x = (ij|b|kl)$$
$$y = (kl|b|ij),$$

we see that each term of this sum contains the factor

$$f(x, y) = (x - y) \ln \frac{x}{y} \qquad (5.5)$$

and that $f(x, y) = 0$ where $x = y$ and is positive otherwise. Hence

$$\frac{d\bar{\bar{H}}}{dt} \leq 0. \qquad (5.6)$$

We have here the Boltzmann form of the H-theorem. Equilibrium corresponds to the situation

$$\frac{dH}{dt} = 0, \tag{5.7}$$

which is satisfied only if

$$x = y. \tag{5.8}$$

11.6 Equilibrium

According to the equilibrium condition just found, we have

$$\ln x = \ln y \tag{6.1}$$

or

$$\ln \frac{v_i v_j (1 \mp v_k)(1 \mp v_l)}{(1 \mp v_i)(1 \mp v_j) v_k v_l} = 0$$

or again

$$f_i + f_j = f_k + f_l. \tag{6.2}$$

We may say that f is conserved in the collision. The other constants of the collision are the energy and momentum, which obey the laws

$$\mathbf{p}_i + \mathbf{p}_j = \mathbf{p}_k + \mathbf{p}_l \tag{6.3}$$

$$\frac{p_i^2}{2m} + \frac{p_j^2}{2m} = \frac{p_k^2}{2m} + \frac{p_l^2}{2m}. \tag{6.4}$$

It may be shown that $f(p)$, the solution of the functional equation (6.2) subject to (6.3) and (6.4), must be of the form

$$f(p) = A + \mathbf{B}\mathbf{p} + C\frac{\mathbf{p}^2}{2m} \tag{6.5}$$

and therefore

$$v = \frac{1}{e^{f(p)} \pm 1} \tag{6.6}$$

or

$$v = \frac{1}{\zeta^{-1} e^{\frac{m}{2}(\mathbf{v} - \mathbf{v}_0)^2 / kT} \pm 1}. \tag{6.7}$$

The distribution depends on the five constants A, $\mathbf{B}$, C or alternatively ζ, $\mathbf{v}_0$, T.

These distributions are steady only if ζ, T, and $\mathbf{v}_0$ are independent of x. Nonuniformities in ζ, T, and $\mathbf{v}_0$ give rise to mass, energy, and momentum transport. An initial inhomogeneous distribution will change to a homogeneous one. Such a redistribution may be studied by Boltzmann's equation, which is the equation of continuity in phase space:

$$\text{div}\,(\rho \mathbf{V}) + \frac{\partial \rho}{\partial t} = C$$

where ρ and $\mathbf{V}$ are density and velocity in phase space. The right hand side C, which is the source term, represents the effect of collisions and is given by equation (4.7). This equation provides the basis for a statistical theory of transport phenomena and a determination of the coefficients of diffusion, thermal conductivity, and viscosity.

The macroscopic description of the irreversible approach to equilibrium is called irreversible thermodynamics.

11.7   Remarks on the *H*-theorem and Boltzmann's Equation

The collision rate entering into Boltzmann's equation, and also into the calculation of $d\bar{\bar{H}}/dt$, is based on a set of simplifying assumptions:
   (a) Only binary collisions are taken into account;
   (b) the walls are ignored;
   (c) the velocity of a molecule is uncorrelated with its position (molecular chaos).

If assumption (c) were not made, we would have a set of coupled equations involving higher order correlation functions.

The *H*-theorem tells us that if (c) is satisfied, then

$$\frac{d\bar{\bar{H}}}{dt} \leq 0 \tag{7.1}$$

and

$$\frac{d\bar{\bar{H}}}{dt} = 0 \tag{7.2}$$

only if the distribution function is F.D. or E.B.

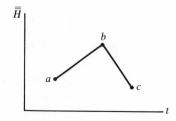

FIGURE 11.1
Local peak in $\bar{\bar{H}}$ at point of molecular chaos.

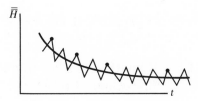

FIGURE 11.2
The approach to equilibrium. The smooth curve is the solution of Boltzmann's equation. The dots are the points at which there is molecular chaos.

The function $\bar{H}$ must of course change discontinuously as the result of individual collisions. Furthermore, if (c) is satisfied, one may see in the following way that $\bar{\bar{H}}$ must be at a local peak [2].

Let molecular chaos be satisfied at $t = t_b$ (Figure 11.1) and let the state of the gas at this time be denoted by B. Then according to the H-theorem

$$\left(\frac{d\bar{\bar{H}}}{dt}\right)_B \le 0 \qquad \text{at } t = t_b^+, \tag{7.3}$$

where t^+ means the derivative on the positive side.

Now consider another state of the gas which is obtained from B by reversing all molecular velocities and is denoted by B'. Assume that B was prepared under initial conditions that are invariant under time reversal. Then the distribution function does not change when $\mathbf{v}$ is replaced by $-\mathbf{v}$. Then B' must have the same value of $\bar{\bar{H}}$ and it must also satisfy the condition of molecular chaos and hence the $\bar{\bar{H}}$-theorem. Therefore

$$\left(\frac{d\bar{\bar{H}}}{dt}\right)_{B'} \le 0 \qquad \text{at } t = t_b^+. \tag{7.4}$$

But the future of B' is the past of B because of the invariance of the equations of motion under time reversal. Hence

$$\left(\frac{d\bar{\bar{H}}}{dt}\right)_B \ge 0 \qquad \text{at } t = t_b^-. \tag{7.5}$$

Hence $\bar{\bar{H}}(t_b)$ is a local peak. Therefore molecular chaos implies that $\bar{\bar{H}}$ is at a local maximum (although not all peaks represent states of molecular chaos).

Figure 11.2 shows the general qualitative behavior of $\bar{\bar{H}}$ for a system that is started off in a nonequilibrium state. $\bar{\bar{H}}$ will tend to decrease until the equilibrium distribution is realized. However, it will fluctuate as it trends downward and even after equilibrium is reached it will continue to fluctuate. Therefore it is certainly not true that $d\bar{\bar{H}}/dt \le 0$ always and consequently the famous paradoxes [3] associated with the H-theorem are not real since they are based on the false assumption that $d\bar{\bar{H}}/dt \le 0$ always.

Since the Boltzmann equation is also based on assumption (c) it cannot be exact either. The solution of the Boltzmann equation is depicted by the smooth curve shown in Figure 11.2. Since (c) is assumed in the derivation of the Boltzmann equation, this smooth curve must try to approximate the dots that represent the points where (c) holds [**4**].

11.8   Remarks on the Quantum Mechanical Boltzmann Problem

The approach to equilibrium (the Boltzmann problem) has just been discussed in μ-space. The Pauli analysis is essentially the same as the classical argument of Boltzmann except for the use of F.D. and E.B. collision rates, therefore the general remarks of Section 11.7 apply to the classical treatment as well.

Let us next consider the approach to equilibrium in Γ-space. This is the fundamental approach and is therefore conceptually simpler, especially if the system is strongly coupled or if there is macroscopic occupation of a single quantum state. It would also be most in accord with the spirit of statistical mechanics to discuss the Boltzmann problem in a general manner without making special assumptions about the dynamics of the system. Unfortunately these (ergodic) investigations have all been so general that their relevance has not been established and progress on the physical side has again been based upon the study of special simple but realistic Hamiltonians. This approach was also initiated by Pauli [**1**] who considered weakly interacting systems and established under certain conditions the so-called master equation

$$\frac{d}{dt} P_n(t) = \sum_m [W_{mn} P_m(t) - W_{nm} P_n(t)] \tag{8.1}$$

where $P_n(t)$ is the probability that the system is in state n at time t and W_{nm} is the transition probability from state n to state m.

It can be shown that solutions to (8.1) approach the desired limit as $t \longrightarrow \infty$ [**5**].

Equations (8.1) and (4.4) are similar in form. They are also valid under similar conditions, namely, when the Hamiltonian may be written in the form $H + U$ where U is a small perturbation. There is also an assumption of randomness needed to establish (8.1) that is similar to Boltzmann's assumption of molecular chaos needed for (4.4). This is Pauli's assumption that the density matrix is diagonal at all times (and not merely at $t = 0$, as assumed in Appendix D). Although van Hove [**6**] showed that Pauli's assumption was not satisfactory, he was able to analyze the same problem by assuming that the density matrix is diagonal at the initial time only, and he was able to derive generalized master equations for the evolution of $P_n(t)$ in time. These equations can be used to investigate the approach to equilibrium.

11.9   Irreversible Thermodynamics [7]

Consider a macroscopic system that has been prepared in a state that is near thermal equilibrium. For example, consider a gas with the distribution function

$$f(x) = \frac{1}{\varsigma^{-1}(x)e^{\frac{1}{2}m[\mathbf{v}-\mathbf{v}_0(x)]^2/kT(x)} \pm 1} \tag{9.1}$$

where the functions $\varsigma(x)$, $T(x)$, and $\mathbf{v}(x)$ depend only slightly on position. This function will not be a solution of Boltzmann's equation. However, it is a possible choice of the distribution function at $t = 0$, and will determine $f(t, x)$ at any later time. We know in fact that $f(t, x)$ will develop into a homogeneous distribution (independent of position) as a consequence of the transport of mass, momentum, and energy. A macroscopic description of these processes is provided by irreversible thermodynamics. Such a description of course depends on more macroscopic variables than thermodynamics, just as the latter itself is based on more variables than hydrodynamics.

Let (A_i) be a set of macroscopic parameters, say pressures and densities in small cells. Let $(\mathring{A}_i)$ represent an equilibrium state and consider a nearby state

$$A_i = \mathring{A}_i + a_i. \tag{9.2}$$

Let us assume that we are close enough to equilibrium so that the entropy may be represented by a quadratic form in the a_i:

$$S = S_0 - \sum_{i,j} g_{ij} a_i a_j. \tag{9.3}$$

Let the rate of entropy change be

$$\frac{dS}{dt} = \sum_i X_i J_i \tag{9.4}$$

where

$$X_i = \frac{\partial S}{\partial a_i} \tag{9.4a}$$

$$J_i = \frac{da_i}{dt}. \tag{9.4b}$$

By the time derivative $\dfrac{da_i}{dt}$, we shall understand

$$\frac{da_i}{dt} \equiv \frac{1}{\tau}[a_i(t + \tau) - a_i(t)] \tag{9.5}$$

where τ is a short time subject to the following conditions:

$$\tau \gg \tau \text{ (collision)}$$

$$\tau \ll \tau \text{ (relaxation)} \tag{9.5a}$$

where τ(collision) is the collision time characteristic of the mechanism that establishes the equilibrium, and τ(relaxation) is the time required for the equilibrium to be attained from the initial nonequilibrium state.

11.10   Ensemble Averages

Assuming a microcanonical ensemble and Boltzmann's relation between the entropy and the probability, we shall write, even for a nonequilibrium situation,

$$S = k \ln W$$

$$W = e^{S/k}$$

$$W = W_0 e^{\Delta S/k} \tag{10.1}$$

where $k \ln W_0 = S_0$ holds at equilibrium and $\Delta S = S - S_0$. Then

$$k \frac{\partial}{\partial a_i} (\ln W) = \frac{\partial}{\partial a_i} (\Delta S). \tag{10.2}$$

Denote an ensemble average by $\langle \ \rangle$. Then by (9.4a) and (10.2)

$$\langle a_m X_n \rangle = \int \cdots \int a_m W \frac{\partial}{\partial a_n} (\Delta S) \, da_1 \ldots$$

$$= k \int \cdots \int a_m \frac{\partial W}{\partial a_n} \, da_1 \ldots$$

$$= -k \delta_{mn}. \tag{10.3}$$

where (10.3) follows from an integration by parts.

11.11   Onsager's Theorem

When deviations from equilibrium are small, it is assumed that there is a linear relation between X_n and J_n:

$$J_n = \sum_p L_{np} X_p. \tag{11.1}$$

By virtue of microscopic reversibility it is now possible to prove that the matrix $\|L_{mn}\|$ is symmetric:

$$L_{mn} = L_{nm}. \tag{11.2}$$

This is Onsager's theorem.

In order to prove this theorem we now formulate microscopic reversibility as follows:

$$\{a_m(t)a_n(t + \tau)\} = \{a_m(t)a_n(t - \tau)\} \tag{11.3}$$

where $\{\ \}$ represents a time average, defined as

$$\{f(t)\} = \lim_{T \to \infty} \frac{1}{2T} \int_{-T}^{T} f(t)\,dt. \tag{11.4}$$

Then

$$\{f(t)\} = \{f(t - \tau)\} \tag{11.5}$$

and

$$\{a_m(t)a_n(t + \tau)\} = \{a_m(t - \tau)a_n(t)\}.$$

Hence equation (11.3) is equivalent to

$$\{a_m(t)a_n(t - \tau)\} = \{a_n(t)a_m(t - \tau)\}, \tag{11.6}$$

i.e., these correlation coefficients are symmetric in their indices. By the ergodic theorem time averages are equal to ensemble averages. Hence

$$\langle a_m(t)a_n(t - \tau)\rangle = \langle a_n(t)a_m(t - \tau)\rangle \tag{11.7}$$

where $\langle\ \rangle$ means ensemble average. One may of course write

$$\langle a_m(t)a_n(t)\rangle = \langle a_n(t)a_m(t)\rangle \tag{11.8}$$

and subtract from (11.7). Then by (9.5) and (9.4b)

$$\langle a_m(t)J_n(t)\rangle = \langle a_n(t)J_m(t)\rangle. \tag{11.9}$$

This last equation is therefore a consequence of microscopic reversibility.

If we now multiply (11.1) by a_m and take the expectation value, we get, according to (10.3),

$$\langle a_m J_n\rangle = \sum_p L_{np}\langle a_m X_p\rangle$$

$$= -kL_{nm}.$$

We may also write

$$\langle a_n J_m\rangle = -kL_{mn}. \tag{11.10}$$

From (11.9) and (11.10), Onsager's theorem (11.2) follows. The J_m and X_n are macroscopic quantities and the L_{mn} are macroscopic coefficients. It has been shown that the matrix $||L_{mn}||$ is symmetric [8].

11.12　Rate of Entropy Production

The second law of thermodynamics may be written in the form

$$dS = \frac{dQ}{T} + d_iS \tag{12.1}$$

where dQ is the heat supplied to the system by its surroundings and d_iS is the entropy produced inside the system by irreversible processes. In ordinary thermostatics, one considers situations in which $d_iS = 0$ or in which $d_iS > 0$, but is otherwise unknown.

In irreversible thermodynamics it is necessary to actually compute the d_iS associated with the approach to equilibrium. To do this, recall that according to Section 11.9 the total volume is divided into small cells which may have different local temperatures. Although this system is not in thermostatic equilibrium, one assumes that the Gibbs relation is still valid for each cell (α):

$$T_\alpha dS_\alpha = dU_\alpha + dW_\alpha - \sum_k \mu_{\alpha k} \, dM_{\alpha k}. \tag{12.2}$$

Then $d_iS = 0$ for each cell but of course $d_iS \neq 0$ for the total system as long as the T_α (for example) are not all the same. The justification of this procedure requires proof that the entropy is well defined under these conditions and that (12.2) is in fact still correct. These points have been investigated by means of kinetic theory as follows.

It is possible to expand the velocity distribution in a series

$$f = f^0 + f^1 + f^2 + \cdots \tag{12.3}$$

in such a way that f^0, the Maxwell-Boltzmann distribution, corresponds to thermostatics, while f^1 corresponds to the formulas of irreversible thermodynamics, and f^2 leads to deviations from these formulas [9].

We next discuss some typical applications of the general method of irreversible thermodynamics [7].

11.13　Thermo-osmosis (Thermo-mechanical Effect)

When two or more irreversible phenomena are occurring simultaneously, there is interference between them. For example, consider a gas in which there is simultaneous energy and mass transport. Then

$$J_1 = L_{11}X_1 + L_{12}X_2$$
$$J_2 = L_{21}X_1 + L_{22}X_2 \tag{13.1}$$

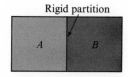

Rigid partition

A | B

FIGURE 11.3
Partition permits energy and mass to pass through it.

where X_1 and X_2 are determined by the temperature and concentration gradients respectively, while J_1 and J_2 are the corresponding energy and mass currents. There is thus an energy current due to the concentration gradient if $L_{12} \neq 0$, and a mass current associated with the temperature gradient if $L_{21} \neq 0$.

Let us consider a container of gas divided into two parts by a partition that permits both energy and mass to flow through it (Figure 11.3). Suppose that the outside wall on the other hand is both adiabatic and impermeable. Let the total system differ from equilibrium by the small amounts $\delta M_A, \delta U_A$. Then

$$\delta M_A + \delta M_B = 0$$
$$\delta U_A + \delta U_B = 0. \tag{13.2}$$

The total entropy change to terms of the second order is

$$\Delta S = \Delta S_A + \Delta S_B$$
$$= [S_{UU}(\Delta U)^2 + 2S_{UM}(\Delta U)(\Delta M) + S_{MM}(\Delta M)^2]_A, \tag{13.3}$$

since terms of the first order cancel and contributions from B and A in the second order are equal. The rate of entropy production is

$$\frac{1}{2}\frac{d(\Delta S)}{dt} = [S_{UU}\,\Delta U + S_{UM}\,\Delta M]\frac{d}{dt}(\Delta U) + [S_{MU}(\Delta U) + S_{MM}(\Delta M)]\frac{d}{dt}(\Delta M)$$

$$\tag{13.4}$$

and therefore

$$X_U = S_{UU}\,\Delta U + S_{UM}\,\Delta M \qquad J_U = \frac{d}{dt}(\Delta U)$$

$$X_M = S_{MU}\,\Delta U + S_{MM}\,\Delta M \qquad J_M = \frac{d}{dt}(\Delta M) \tag{13.5}$$

so that

$$X_U = \Delta S_U = \Delta\left(\frac{1}{T}\right) \tag{13.6a}$$

$$X_M = \Delta S_M = \Delta\left(\frac{-\mu}{T}\right). \tag{13.6b}$$

With the aid of these results it is possible to establish a connection between two kinds of experiments, which are distinguished as follows:

(a) no temperature gradient;

(b) no mass current.

In the situation (a),

$$\Delta T = 0, \qquad X_U = 0 \qquad (13.7a)$$

and therefore by (13.1),

$$\left(\frac{J_U}{J_M}\right)_{\Delta T=0} = \frac{L_{UM}}{L_{MM}}. \qquad (13.7b)$$

On the other hand, under conditions (b),

$$0 = L_{MU} X_U + L_{MM} X_M \qquad (13.8a)$$

$$\frac{L_{MU}}{L_{MM}} = -\left(\frac{X_M}{X_U}\right)_{J_M=0}. \qquad (13.8b)$$

The ratio of currents measured in (a) would be unrelated to the ratio of forces measured in (b) if the matrix $\|L_{mn}\|$ were arbitrary. Because of microscopic reversibility and the Onsager result, however, we have $L_{UM} = L_{MU}$ and therefore

$$-\left(\frac{X_M}{X_U}\right)_{J_M=0} = \left(\frac{J_U}{J_M}\right)_{\Delta T=0}$$

or

$$\left(\frac{J_U}{J_M}\right)_{\Delta T=0} = \left[\frac{\Delta(\mu/T)}{\Delta(1/T)}\right]_{J_M=0} = \mu - T\left(\frac{\partial \mu}{\partial T}\right)_{J_M=0}$$

$$= h - TV\left(\frac{\partial P}{\partial T}\right)_{J_M=0}. \qquad (13.9)$$

The left and right sides of this equation are measurable in independent experiments.

11.14   Thermoelectric Effect

The theory of the thermocouple was first worked out by Lord Kelvin. This first treatment was followed by many others but the theory did not become clear until it was recognized that all derivations depended in some way on the principle of microscopic reversibility.

Consider a wire in which there are simultaneously electrical and heat currents. Then

$$J_1 = L_{11} X_1 + L_{12} X_2$$
$$J_2 = L_{21} X_1 + L_{22} X_2 \qquad (14.1)$$

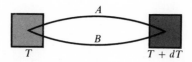

FIGURE 11.4 Thermocouple.

where X_1 and X_2 are proportional to the temperature and voltage drops respectively. J_1 and J_2 are the thermal and electrical currents respectively. There is an electrical current due to the temperature difference (L_{21}) as well as a thermal current due to the voltage drop (L_{12}).

Let us consider in particular a thermocouple, consisting of metals A and B joined at differing temperatures, T, and $T + dT$ (Figure 11.4). Besides the Joule heating there are the following three effects.

(1) Seebeck effect, a thermal emf between the two junctions ($d\varepsilon$).

(2) Peltier effect, a cooling or heating of a junction by the current. The rate at which Peltier heat is absorbed is proportional to the current and changes sign if the current is reversed: it is ΠI where Π is the Peltier coefficient and I is the electric current.

(3) Thomson effect, a heating or cooling of a homogeneous conductor in which there is a temperature gradient. This is also proportional to the current and reverses sign with it: it is $\sigma I \, dT/dx$ where σ is the Thomson coefficient.

To analyze the thermoelectric effect consider a thermocouple in one leg of which a condenser is inserted (Figure 11.5). Let the potential difference across the condenser be $\Delta\varepsilon$. Let the condenser have no heat capacity. The change in entropy of the total system is the sum of the entropy change in its parts:

$$dS = dS_1 + dS_2 + dS_3. \tag{14.2}$$

Suppose the energy dU leaves heat reservoir 2 and arrives at 1, and that this results in an additional charge dq on the condenser. Then

$$dS = \frac{dU}{T_1} - \frac{dU}{T_2} - \frac{(\Delta\varepsilon) \, dq}{T_3}$$

$$\frac{dS}{dt} = \frac{dU}{dt}\left(\frac{1}{T_1} - \frac{1}{T_2}\right) - \frac{dq}{dt}\frac{\Delta\varepsilon}{T_3}$$

$$= \frac{dU}{dt}\frac{\Delta T}{T^2} - \frac{dq}{dt}\frac{\Delta\varepsilon}{T}. \tag{14.3}$$

The fluxes are

$$J = \frac{dU}{dt}$$

$$I = \frac{dq}{dt}. \tag{14.4}$$

FIGURE 11.5
Thermocouple with condenser in one leg.

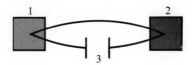

The forces are

$$\frac{\Delta T}{T^2} \quad \text{and} \quad -\frac{\Delta \varepsilon}{T}. \tag{14.5}$$

Then

$$I = L_{11}\left(-\frac{\Delta \varepsilon}{T}\right) + L_{12}\left(\frac{\Delta T}{T^2}\right) \tag{14.6a}$$

$$J = L_{21}\left(-\frac{\Delta \varepsilon}{T}\right) + L_{22}\left(\frac{\Delta T}{T^2}\right). \tag{14.6b}$$

Consider first the case in which $I = 0$. Then we obtain the Seebeck coefficient:

$$\frac{d\varepsilon}{dT} = \frac{L_{12}}{L_{11}}\frac{1}{T}. \tag{14.7}$$

Next if $\Delta T = 0$, we find the Peltier coefficient:

$$\Pi = \frac{J}{I} = \frac{L_{21}}{L_{11}}. \tag{14.8}$$

By Onsager's result the Peltier and Seebeck effects are related as follows:

$$\frac{d\varepsilon}{dT} = \frac{\Pi}{T}. \tag{14.9}$$

This is one of Kelvin's relations. The other relation is just a statement of the first law and was never controversial: namely,

$$\frac{d\Pi}{dT} = \frac{d\varepsilon}{dT} + \sigma_B - \sigma_A \tag{14.10}$$

where $\sigma I\, dT/dx$ is the Thomson heat and σ of course is different for metals A and B.

11.15   Arbitrariness in Choice of Forces

We have by (9.4) and (11.1)

$$\frac{d(\Delta S)}{dt} = \sum_{m,n} L_{mn}\, X_m\, X_n. \tag{15.1}$$

Although there is obviously some arbitrariness in the splitting of $d/dt\,(\Delta S)$ into forces and fluxes, we of course require that $d/dt\,(\Delta S)$ be the same for all such choices.

Let X_m and X'_m be different choices of forces such that

$$X'_m = \sum_n R_{mn} X_n \tag{15.2}$$

or in matrix form

$$X' = R\,X. \tag{15.2a}$$

In matrix form, equation (15.1) is

$$\frac{d(\Delta S)}{dt} = X^{\mathrm{T}} L X \tag{15.1a}$$

where T means transposed. Then

$$\frac{d(\Delta S)}{dt} = X'^{\mathrm{T}}\,L'X' = X^{\mathrm{T}}(R^{\mathrm{T}}L'R)X$$

or

$$R^{\mathrm{T}}L'R = L \tag{15.3}$$

and

$$R^{\mathrm{T}}(L')^{\mathrm{T}}R = L^{\mathrm{T}}. \tag{15.4}$$

Therefore, if L' is symmetric, L is also symmetric.

11.16   Positive Definiteness of Entropy Production

In the present analysis we assume that the rate of entropy production is never negative. Therefore the quadratic form (15.1) must be positive definite. Hence

$$L_{11} > 0 \qquad L_{22} > 0 \qquad |L_{mn}| > 0 \tag{16.1}$$

with an obvious generalization to the case of several variables. These conditions are restrictions on the transport phenomena and are analogous to the conditions for thermodynamic stability derived earlier. Just as the stability conditions imply that certain parameters such as specific heat are positive, the above conditions imply that other empirical parameters such as the thermal conductivity are positive.

Suppose that J_1 and J_2 refer to energy and mass currents again, and let

$$J_2 = 0. \tag{16.2}$$

Then

$$J_1 = L_{11}X_1 + L_{12}\left(-\frac{L_{21}X_1}{L_{22}}\right) = \frac{|L_{mn}|}{L_{22}} X_1 \tag{16.3}$$

or

$$J_1 = \frac{|L_{mn}|}{L_{22}}\left(-\frac{\Delta T}{T^2}\right)$$

$$J_1 = \frac{|L_{mn}|}{L_{22}}\Delta(\ln \chi). \tag{16.4}$$

Hence the coefficient of thermal conductivity is positive:

$$\lambda = \frac{|L_{mn}|}{L_{22}}\frac{1}{T^2} > 0. \tag{16.5}$$

Similarly,

$$J_2 = \frac{|L_{mn}|}{L_{11}}X_2$$

$$= -\frac{|L_{mn}|}{L_{11}}\Delta(\mu/T)$$

$$= -\frac{|L_{mn}|}{L_{11}}\Delta(\ln \zeta). \tag{16.6}$$

Equation (16.6) corresponds to (16.4); the sign of the diffusion coefficient is fixed in the same way as that of the thermal conductivity.

Notes and References

1. W. Pauli, *Sommerfeld Festschrift*, p. 30 (Hirzel, Leipzig, 1928). See also, R. C. Tolman, *The Principles of Statistical Mechanics* (Clarendon Press, Oxford, 1938).
2. K. Huang, *Statistical Mechanics*, p. 85 (Wiley, New York, 1963). (This argument is due to F. E. Low.)
3. Reversal paradox: The $\overline{\overline{H}}$-theorem is inconsistent with time reversal invariance.

   Recurrence paradox: The $\overline{\overline{H}}$-theorem is inconsistent with Poincaré's theorem according to which a system (of finite energy and volume) will, after a sufficiently long time, return to an arbitrarily small neighborhood of almost any given state.

   Both of these apparent paradoxes result from the false premise that $d\overline{\overline{H}}/dt$ is never positive.

4. K. Huang, *Statistical Mechanics*, Chapter 4 (Wiley, New York, 1963).

5. W. Feller, *An Introduction to Probability Theory and Its Applications*, Vol. 1, Chapter 15 (Wiley, New York, 1950). The master equation is also known as the Chapman-Kolmogorov equation.

6. L. van Hove, *Physica* 21: 517 (1955); *Physica* 23: 441 (1957).

7. S. R. de Groot, *Thermodynamics of Irreversible Processes* (Interscience, New York, 1952).

8. The Kubo formula for the conductivity tensor provides a quantum mechanical illustration of this result:

$$j_\mu = \sigma_{\mu\nu} E_\nu$$

$$\sigma_{\mu\nu} = \sigma_{\nu\mu} = \left(\frac{1}{2kT}\right) \int_{-\infty}^{+\infty} dt \, \mathrm{Tr}[\rho \, J_\mu(t) \, J_\nu(0)].$$

R. Kubo, *Can. J. Phys.* 34: 1274 (1956).

9. See reference [7]. This investigation was made by Prigogine.

Analysis of the Kelvin and Clausius Formulations

Schematic Representation of Kelvin and Clausius Statements

A physical system working in a cycle so that it converts heat to work, or vice versa, is shown schematically in Figure A.1. Here $t_2 > t_1$, as determined

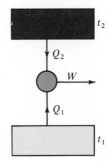

FIGURE A.1
Physical system working in cycle between temperatures t_1 and t_2

by the direction of spontaneous heat flow. Let Q_2 and Q_1 be the heat absorbed by the system at the two temperatures.

The Kelvin and Clausius statements tell us that

$$W > 0 \quad \text{implies} \quad Q_1 < 0 \qquad \text{(K)}$$

$$Q_1 > 0 \quad \text{implies} \quad W < 0. \qquad \text{(C)}$$

(The cyclic system absorbs or rejects heat at the lower temperature according to whether $Q_1 > 0$ or $Q_1 < 0$, respectively. By the first law

$$Q_1 + Q_2 = W. \qquad \text{(I)}$$

[233]

Therefore

$$Q_2 < 0 \quad \text{if} \quad Q_1 > 0 \quad \text{by (C) and (I)},$$
$$Q_2 > 0 \quad \text{if} \quad W > 0 \quad \text{by (K) and (I)}.$$

It is easy to show that statements (K) and (C) are equivalent [1]. They may also both be expressed as follows

$$WQ_1 < 0. \tag{II}$$

Reversible System

We shall now consider a physical system operating in a Carnot cycle and shall show that no system can be more efficient than a reversible one.

We begin by assuming the contrary, namely, that there is an irreversible system with greater efficiency (η') than that of the reversible system (η). Let the irreversible system be used to drive the reversible one (Figure A.2). (The

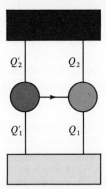

FIGURE A.2
Irreversible system driving reversible system.

two systems may be matched to arbitrary accuracy by running them through many cycles). Then

$$Q_2' + Q_1' = W > 0$$

and

$$Q_2 + Q_1 = -W,$$

since

$$Q_2 + Q_1 + Q_2' + Q_1' = 0. \tag{I'}$$

For the efficiencies we have

$$\eta' = \frac{W}{Q_2'}$$

$$\eta = \frac{W}{-Q_2}.$$

Then by hypothesis

$$\eta' > \eta \tag{A.1}$$

or

$$\frac{W}{Q_2'} > -\frac{W}{Q_2}$$

or

$$Q_2' < -Q_2.$$

Hence

$$Q_2' + Q_2 < 0.$$

By (I)'

$$Q_1' + Q_1 > 0.$$

Therefore the composite system violates the Clausius statement. Hence we reject (A.1) and conclude

$$\eta' \le \eta. \tag{A.2}$$

Consider now any two reversible systems, A and B. First driving B with A, and then A with B, we have

$$\eta_A \le \eta_B$$

and

$$\eta_B \le \eta_A$$

or

$$\eta_A = \eta_B. \tag{A.3}$$

Hence all reversible engines—independent of their physical construction—have the same efficiency.

Absolute Temperature Scale

It has been shown that η_R is independent of the physical system working in the reversible Carnot cycle and depends only on t_1 and t_2. We may therefore set up an absolute temperature scale by using a reversible engine as a thermometer. An absolute temperature may then be defined as follows:

$$\frac{T_1}{T_2} = 1 - \eta_R(t_1, t_2). \tag{A.4}$$

This equation may also be written

$$\frac{T_1}{T_2} = 1 - \frac{W}{Q_2} = -\frac{Q_1}{Q_2}$$

or

$$\frac{Q_1}{T_1} + \frac{Q_2}{T_2} = 0,$$

and

$$\oint \frac{dQ}{T} = 0 \qquad (A.5)$$

for a reversible Carnot cycle. By paving the area bounded by an arbitrary closed path with Carnot cycles, equation (A.5) may again be established generally. Hence the temperature defined in terms of η_R, the efficiency of a reversible engine, is an integrating denominator for dQ and therefore the same as the absolute temperature previously defined.

If a physical system is carried through a Carnot cycle in an unspecified way (either reversibly or irreversibly) we have

$$\eta \le \eta_R \qquad (A.6)$$

$$\frac{Q_1}{T_1} + \frac{Q_2}{T_2} \le 0$$

and

$$\oint \frac{dQ}{T} \le 0. \qquad (A.7)$$

The equality sign holds if the cycle is reversible.

We have already seen that if

$$\Delta S_A > \frac{\Delta Q_A}{T},$$

there is no way of restoring both the system and its environment to their initial states.

According to (7.2) of Chapter 2,

$$\Delta S = \int \frac{dQ}{T} + \Delta S_i.$$

For cyclic systems,

$$\Delta S = 0,$$

and therefore

$$\oint \frac{dQ}{T} = -\Delta S_i$$

or

$$\oint \frac{dQ}{T} = 0 \qquad \text{for quasi-static processes} \qquad (A.8)$$

and

$$\oint \frac{dQ}{T} < 0 \qquad \text{for nonquasi-static processes} \qquad (A.9)$$

since ΔS_i is always positive. Therefore the Kelvin and Clausius statements (A.7) agree with our previous statement (7.2) of Chapter 2.

Reference

1. The analysis of the Kelvin and Clausius formulations described here follows that of E. Fermi, *Thermodynamics* (Dover, New York, 1956).

Collection of Thermodynamic Formulas

Including Q and W we have 10 thermodynamic variables. The number of partial derivatives $(\partial x_i/\partial x_j)_{x_k}$ is $10 \cdot 9 \cdot 8 = 720$. Between any four of these 720 derivatives there is a relation, and hence $C_4{}^{720} \sim 10^{10}$ relations altogether. Bridgman has shown how these relations may be summarized by tabulating $(\partial x_i/\partial x_j)_{x_k}$ in terms of some standard set of three, for example, $(\partial V/\partial T)_P$, $(\partial V/\partial P)_T$, $(\partial Q/\partial T)_P$. Then one may systematically obtain a relation among any four given $(\partial x_i/\partial x_j)_{x_k}$ by eliminating the standard three in terms of which they are expressed. The classification can be further simplified by writing

$$\left(\frac{\partial x_i}{\partial x_j}\right)_{x_k} = \frac{(\partial x_i/\partial \alpha)_{x_k}}{(\partial x_j/\partial \alpha)_{x_k}}$$

and tabulating only the $(\partial x_i/\partial \alpha)_{x_k}$. It is then sufficient to tabulate the 90 derivatives $(\partial x_i/\partial \alpha)_{x_k}$ in 10 groups of 9. Each group is characterized by an α and an x_k. Of course it is not important how α is chosen and one sometimes writes $(\partial x_i/\partial \alpha)_{x_k} = (\partial x_i)_k$.

London Rigidity

There is a simple way to look at equation (9.5) of Chapter 4, which was also found by London. The physical idea is that the superconductive state is very stable or "rigid" with respect to perturbations caused by a weak magnetic field. The idea may be explained as follows.

The connection between the velocity of an electron and its canonical momentum in an electromagnetic field is

$$\mathbf{v} = \frac{d\mathbf{x}}{dt} = \frac{1}{m}\left(\mathbf{p} + \frac{e}{c}\mathbf{A}\right). \tag{C.1}$$

The corresponding quantum mechanical operator for the total electron current is then

$$\mathbf{j}(\mathbf{x}) = \sum_{i=1}^{n} \mathbf{j}(\mathbf{x}_i)\, \delta(\mathbf{x} - \mathbf{x}_i) \tag{C.2}$$

where

$$\mathbf{j}(\mathbf{x}_i) = -e\mathbf{v}_i = -\frac{e}{m}\left(\mathbf{p} + \frac{e}{c}\mathbf{A}\right)_i, \tag{C.3}$$

so that

$$j(x) = \overset{\circ}{j}(x) - \frac{e^2}{mc}\sum_{i=1}^{n} A(x_i)\, \delta(x - x_i) \tag{C.4}$$

where $\overset{\circ}{j}(x)$ is the current operator when $\mathbf{A}$ vanishes.

The expectation value of the current in the superconducting state $\Psi_S(= |S\rangle)$ subject to a vector potential $(\mathbf{A})$ is then

$$\langle SA| j(\mathbf{x}) |SA\rangle$$
$$= \int \ldots \int \Psi^*_{SA}(\mathbf{x}_1, \ldots, \mathbf{x}_n) j(\mathbf{x}, \mathbf{x}_1, \ldots, \mathbf{x}_n) \Psi_{SA}(\mathbf{x}_1, \ldots, \mathbf{x}_n)\, d\mathbf{x}_1, \ldots, d\mathbf{x}_n$$

Let the wave function be denoted by $|S, 0\rangle$ when $\mathbf{A} = 0$. For this state there is no current:

$$\langle SO|\overset{\circ}{\mathbf{j}}(\mathbf{x})|SO\rangle \equiv 0. \tag{C.6}$$

London then makes the assumption that the superconduting state is "rigid," or that $|SO\rangle$ and $|SA\rangle$ are not very different in the sense that

$$\langle SA|\overset{\circ}{\mathbf{j}}(\mathbf{x})|SA\rangle \cong \langle SO|\overset{\circ}{\mathbf{j}}(\mathbf{x})|SO\rangle = 0, \tag{C.7}$$

as long as the perturbing magnetic field associated with $\mathbf{A}$ is not too strong and does not vary too rapidly in space.

It then follows from (C.5) and (C.4) that

$$\langle S|\mathbf{j}(\mathbf{x})|S\rangle_A = \langle SA|\overset{\circ}{\mathbf{j}}(\mathbf{x})|SA\rangle - \frac{e^2}{mc}\left\langle SA\left|\sum_i A(\mathbf{x}_i)\,\delta(\mathbf{x} - \mathbf{x}_i)\right|SA\right\rangle \tag{C.8}$$

and by (C.7) the first term vanishes.

Therefore

$$\langle S|\mathbf{j}(\mathbf{x})|S\rangle_A = -\frac{e^2}{mc}n_s(\mathbf{x})\mathbf{A}(\mathbf{x}) \tag{C.9}$$

where

$$n_s(\mathbf{x}) = \sum_i \int \ldots \int |\Psi_{SA}(\mathbf{x}_1, \ldots, \mathbf{x}_n)|^2\,\delta(\mathbf{x} - \mathbf{x}_i)\,d\mathbf{x}_1, \ldots, d\mathbf{x}_n. \tag{C.10}$$

This is equation (9.5) of Chapter 4 where

$$\Lambda = \frac{m}{n_s e^2}$$

and n_s is the number of electrons condensed into the superconducting state.

From this viewpoint the essential feature of the superconducting state appears to be its stability with respect to a weak magnetic perturbation, say V. A criterion for this condition is that the following sum be very small:

$$\sum_\alpha \left|\frac{\langle \alpha|V|S\rangle}{E_\alpha - E_S}\right|^2$$

according to first order perturbation theory. From this equation it would follow that the superconductive state is stable if there are no other states with energies close to E_S (energy gap) or if the matrix elements connecting with these states are very small.

The energy gap is from this point of view not a necessary condition in principle. It is certainly not a sufficient condition, since insulators are also characterized by an energy gap. In this case, (C.7) is not satisfied:

$$\langle SA|\overset{\circ}{\mathbf{j}}|SA\rangle \neq \langle SO|\overset{\circ}{\mathbf{j}}|SO\rangle$$

and therefore the first term of (C.8) does not vanish. One may show in fact that in the case of insulators it cancels the second term [1].

Reference

1. J. R. Schrieffer, *Theory of Superconductivity*, Section 8.1 (Benjamin, New York, 1964).

Time Dependence of $\overline{H}$ in Quantum Theory [1]

The time dependence of ρ is given by equation (21.11) of Chapter 6 or by its formal solution

$$\rho(t) = U(t)\rho(0)U^{-1}(t) \tag{D.1}$$

where $U(t)$ is a unitary matrix. In component form we have

$$\rho_{kl}(t) = \sum_{m,n} U_{km}(t)\rho_{mn}(0)U_{nl}{}^{-1}(t)$$

$$= \sum_{m,n} U_{km}(t)U^*{}_{ln}(t)\rho_{mn}(0). \tag{D.2}$$

We shall be interested in the case for which $\rho(0)$ is diagonal. Then

$$\rho_{kk}(t) = \sum_{m} |U_{km}(t)|^2\rho_{mm}(0). \tag{D.3}$$

Again by unitarity

$$\sum_{k} |U_{kn}(t)|^2 = \sum_{n} |U_{kn}(t)|^2 = 1. \tag{D.4}$$

One now makes use of the following condition which is familiar from the classical proof of the H-theorem:

$$\begin{array}{ll} f(x, y) > 0 & x \neq y \\ = 0 & x = y \end{array} \tag{D.5}$$

where

$$f(x, y) = x \ln \frac{x}{y} - x + y.$$

Now let

$$x = \rho_{kk}(0)$$

$$y = \rho_{nn}(t).$$

[242]

Then

$$Q_{kn} \equiv f[\rho_{kk}(0), \rho_{nn}(t)] \geq 0 \qquad \text{(D.6)}$$

and therefore

$$\sum_{k,n} |U_{nk}|^2 Q_{kn} \geq 0. \qquad \text{(D.7)}$$

This last equation may be written out in detail:

$$\sum_{n,k} |U_{nk}(t)|^2 [\rho_{kk}(0) \ln \rho_{kk}(0) - \rho_{kk}(0) \ln \rho_{nn}(t) - \rho_{kk}(0) + \rho_{nn}(t)] \geq 0. \quad \text{(D.8)}$$

The four terms may be evaluated as follows, by (D.4):

$$\sum_{k,n} |U_{nk}(t)|^2 \rho_{kk}(0) \ln \rho_{kk}(0) = \sum_{k} \rho_{kk}(0) \ln \rho_{kk}(0) \sum_{n} |U_{nk}(t)|^2$$

$$= \sum_{k} \rho_{kk}(0) \ln \rho_{kk}(0).$$

By (D.3):

$$\sum_{k,n} |U_{nk}(t)|^2 \rho_{kk}(0) \ln \rho_{nn}(t) = \sum_{n} \ln \rho_{nn}(t) \sum_{k} |U_{nk}(t)|^2 \rho_{kk}(0)$$

$$= \sum_{n} \rho_{nn}(t) \ln \rho_{nn}(t).$$

Again by (D.4) and the usual normalization of the trace:

$$\sum_{k,n} |U_{nk}(t)|^2 \rho_{kk}(0) = \sum_{k} \rho_{kk}(0) = 1$$

$$\sum_{n,k} |U_{nk}(t)|^2 \rho_{nn}(t) = \sum_{n} \rho_{nn}(t) = 1.$$

Therefore (D.8) becomes

$$\sum_{n} \rho_{nn}(0) \ln \rho_{nn}(0) \geq \sum_{n} \rho_{nn}(t) \ln \rho_{nn}(t) \qquad \text{(D.9)}$$

or

$$\bar{H}(0) \geq \bar{H}(t).$$

The essential input needed for the proof is the assumption

$$\rho_{k1}(0) = 0 \qquad k \neq 1$$

This assumption in turn may be related to an hypothesis about random phases in the quantum ensemble. We have

$$\rho(k, 1) = \sum_{\alpha} \langle k | \alpha \rangle W_{\alpha} \langle \alpha | 1 \rangle.$$

Let

$$\langle \alpha | 1 \rangle = \gamma_1(\alpha) e^{i\phi_1(\alpha)}.$$

Then

$$\rho(k, 1) = \sum_\alpha W\alpha\gamma_1(\alpha)\gamma_k(\alpha)e^{i[\phi_1(\alpha)-\phi_k(\alpha)]}$$

$$= \sum_\alpha W\alpha\gamma_1(\alpha)\gamma_k(\alpha)[\cos(\phi_1(\alpha) - \phi_k(\alpha)) + i\sin(\phi_1(\alpha) - \phi_k(\alpha))].$$

The assumption of incomplete knowledge about the system may now be formulated in such a way as to imply that there are no special phase relations between the $\langle \alpha | 1 \rangle$ and so that

$$\rho(k, 1) = 0 \quad \text{if} \quad k \neq 1.$$

That is, one makes two assumptions about the quantum ensemble: (1) equal a priori probabilities, and (2) random a priori phases.

Reference

1. R. C. Tolman, *The Principles of Statistical Mechanics* (Clarendon Press Oxford, 1938).

Index